Praise for
Managerial Analytics

"As the new era with abundant data sources is upon us, books on data analytics are gaining traction. Most of them either focus on the technical side of analytics or on the underlying business processes and the business of analytics. A manager who wants to learn the underlying techniques and their applicability is bound to tutorials from a data scientist with good communication skills. This book gives managers the opportunity to learn the concepts by themselves, and thus, it should be on a bookshelf of everyone who leads and manages analytics efforts. The book covers the most important methodologies and concepts in data analytics from a non-technical perspective. Each methodology is nicely wrapped with examples and use cases, and it does not require technical knowledge. If I were to venture in the field of analytics from the business perspective, this would be the first book to read in the morning."

—**Diego Klabjan,** Professor of Industrial Engineering and Management Sciences, Director of Master of Science in Analytics Program, Northwestern University

"An excellent introduction and overview of the field of analytics, *Managerial Analytics* is easily accessible for those new to the field and provides a useful framework for readers with a deeper background. Written from a practitioner's point of view, the book is well stocked with concise and relevant examples. The authors set out to define the boundaries of what is currently possible with analytics tools, and they guide the reader in asking good questions, avoiding common pitfalls, and identifying hidden assumptions when managing their own analytics projects."

—**Michael Freimer, Ph.D.,** Chief Scientist, DemandSignal

"The term 'analytics' means multiple things to different audiences. Watson and Nelson help to bridge the gap between the marketing hype and the technical details, making it easier to evaluate analytics-based solutions and better understand their potential."

—**Irv Lustig, Ph.D.,** Manager, Optimization and Mathematical Software, Business Analytics and Math Science, IBM Research at IBM

Managerial Analytics

Managerial Analytics

An Applied Guide to Principles, Methods, Tools, and Best Practices

Michael Watson
Derek Nelson

Vice President, Publisher: Tim Moore
Associate Publisher and Director of Marketing: Amy Neidlinger
Executive Editor: Jeanne Glasser Levine
Operations Manager: Jodi Kemper
Cover Designer: Alan Clements
Managing Editor: Kristy Hart
Project Editor: Elaine Wiley
Copy Editor: Kitty Wilson
Proofreader: Jess DeGabriele
Indexer: Lisa Stumpf
Senior Compositor: Gloria Schurick
Manufacturing Buyer: Dan Uhrig

© 2014 by Michael Watson and Derek Nelson
Publishing as Pearson
Upper Saddle River, New Jersey 07458

For information about buying this title in bulk quantities, or for special sales opportunities (which may include electronic versions; custom cover designs; and content particular to your business, training goals, marketing focus, or branding interests), please contact our corporate sales department at corpsales@pearsoned.com or (800) 382-3419.

For government sales inquiries, please contact governmentsales@pearsoned.com.

For questions about sales outside the U.S., please contact international@pearsoned.com.

Company and product names mentioned herein are the trademarks or registered trademarks of their respective owners.

Printed in the United States of America

First Printing December 2013

ISBN-10: 0-13-340742-X
ISBN-13: 978-0-13-340742-6

Pearson Education LTD.
Pearson Education Australia PTY, Limited.
Pearson Education Singapore, Pte. Ltd.
Pearson Education Asia, Ltd.
Pearson Education Canada, Ltd.
Pearson Educación de Mexico, S.A. de C.V.
Pearson Education—Japan
Pearson Education Malaysia, Pte. Ltd.

Library of Congress Control Number: 2013950966

To my wife, Kristen,
for all her support throughout this project
MSW

To my wife, Bridget,
who dreamed as a little girl of having a book about
analytics dedicated to her ☺
DKN

Contents

Acknowledgments

This book, covering such a wide subject matter, would not have been possible without the many people we've learned so much from over the years.

While it is impossible to thank everyone, we apologize in advance if we've missed someone.

First, we would like to thank Pete Cacioppi for his help with early versions of the book and suggested text for certain optimization sections. Thanks also to Sara Lewis for reviewing and commenting on the pre-published book.

We'd like to thank all the experts who answered our questions and helped teach us about their specialty. Thank you Irv Lustig (for teaching us how to teach optimization and helping us understand the breadth of analytics), Richard Whisner and Patricio Cofre (for teaching us data modeling and BI systems), Dave Vander Veen (for teaching us about implementing full-scale analytics systems), David and Edith Simchi-Levi (for teaching us about rigorous analysis and how to commercialize this technology), Mark Daskin (for teaching us about optimization and how to present analytics results) and Diego Klabjan, the founding director of Northwestern University's Master of Science in Analytics, (for introducing us to many new ideas in the field of analytics programming).

We would like to thank the leading experts in the field who helped create the analytics movement. We've included an unofficial bibliography with some of the great books and articles we read. We also reference a lot of nice articles in the endnotes. And, we'll keep this online at the book's website (ManagerialAnalytics.com). Hopefully, this list will allow you to further explore topics of interest to you.

Finally, we couldn't have completed the book without the assistance and expert guidance of Jeanne Glasser Levine, our editor at Pearson.

About the Authors

Michael Watson is currently a partner at Opex Analytics and an Adjunct Professor at Northwestern University. At Opex Analytics he helps bring new analytics solutions to companies. Prior to Opex Analytics, he was a manager at IBM in the ILOG supply chain and optimization group. At Northwestern, he teaches a program on operations management and managerial analytics in the McCormick School of Engineering's Masters in Engineering Management (MEM). He teaches optimization in Northwestern's Master of Science in Analytics program. He holds an M.S. and Ph.D. from Northwestern University in Industrial Engineering and Management Sciences.

Derek Nelson is currently a senior principal at OPS Rules and an Adjunct Professor at Northwestern University. At OPS Rules, Derek leverages analytics to help companies improve operational performance. Prior to OPS Rules, Derek held consulting, product management, and technical sales roles in optimization and supply chain software for LogicTools, ILOG, and IBM. At Northwestern, Derek has taught service operations management to undergraduates in the Industrial Engineering and Management Sciences department and will soon be teaching in the Master in Engineering Management (MEM) program. Derek holds an M.S. in Operations Research from Cornell University.

Preface

Upon first hearing the term *analytics* used in common vernacular, we the authors were unimpressed. After all, we had spent our careers doing analytics. Our undergraduate and graduate degrees focused on topics like mathematical optimization, probability, and statistics. In our careers, we analyzed data all day long: creating databases and reporting systems, working with a variety of tools (including spreadsheets, databases, reporting tools, statistical tools, optimization tools, and specialized engineering tools) to test, manipulate, and understand data in meaningful ways that would allow our company and others we worked with to make better decisions about their business.

We never labeled what we did as "analytics," but surely if anyone was doing analytics, we were.

The term's popularity continued to rise. Soon, it seemed like the term had caught fire. People were using it frequently, it was popping up in marketing campaigns, and whole companies were relabeling themselves as "analytics" specialists. Admittedly, we fell into a common trap of initially resisting the term, basically taking the position that we were seeing yet another example of "Everything old is new again." Don't be fooled, we thought: Analytics is simply what we old-timers call math.

Soon, however, it became clear that this was not just a new term but a movement. And it quickly overwhelmed our ability to resist it. We would see a software vendor say that if you implemented that company's solution, you would be doing analytics. We read about firms in Silicon Valley saying that every firm had to be doing analytics or they would not succeed.

As we read more articles and talked to more people, we realized that the term *analytics* was being used in very different ways by different people. We also saw that software companies and certain types of products seemed to hijack the term and use it as if anything outside of what they were doing wasn't analytics. This didn't sit well with us professionally. Even though we knew that what we were doing was analytics, we feared that people wouldn't recognize it as such because so many were using the term inconsistently.

Eventually, we took a step back, got humble, and admitted to ourselves that perhaps there was more to the story than just the topics in which we had deep experience. There *was* something new about analytics. But before we started researching this topic, we couldn't place our finger on what exactly that was.

In other words, there was not a good definition and description of the field of analytics. If someone wanted to know what this analytics movement was about, it was hard to find a complete answer. To make matters more confusing, the term *Big Data* quickly came on the scene and seemed to be used as a synonym for *analytics*.

We figured that if we were confused, others would be as well. And if managers were now being asked to do more with analytics and Big Data, they needed to truly understand what that was. So we decided to write this book to help managers and analysts everywhere better understand the analytics movement.

We wrote this book for people interested in learning what the analytics movement is all about. We wrote it for people who know that their organization needs analytics to improve but needs to cut through the hype, find a clear definition, and better implement analytics solutions. We wrote it for consultants and software providers so they can deliver better results for their clients. We wrote it for people who are tasked with evaluating analytics solutions but may not know where to start asking questions. We want to see more people, in all kinds of organizations, doing better analytics. We wrote the book from the viewpoint of what a manager needs to know about analytics. Since analytics is technical, we do go into some technical details. But we do so only to help managers gain intuition and insight. And the field is much too large to cover everything in a book like this. We had to pick and choose which topics to cover and to what depth. Certainly, this list could have been different. But, hopefully, we have provided enough information and references that you will be in a better position to explore topics that are of more interest to you.

We hope that a lot of different people get value from this book. We hope managers will be in a better position to evaluate and run projects. We hope that specialists in certain areas will better see how their specialty fits in with other specialties. And we hope that this book will inspire or give you some good ideas to make a significant difference in your organization. Done well, analytics can do everything from help companies create new strategies and save money to help healthcare organizations save lives.

Part I
Overview

1

What Is Managerial Analytics?

Confusion About the Meaning of Analytics

The field of analytics (and its frequently used synonym *Big Data*) has captured the imagination of managers everywhere.

Companies often publicly claim that they are committed to analytics. General Electric recently announced a big push into "analytics" to take advantage of the data generated by its own industrial machines.[1] IBM and Ohio State University announced the creation of an analytics center that is projected to employ up to 500 people doing analytics work. Google touts its own internal analytics for its superior searching capabilities and offers its services to others to track their own websites. This list of companies and their newly publicized analytics capability could go on and on.

Not wanting to miss the excitement, consulting and software companies are also touting their analytics services and products. The more skeptical may think that some of these firms didn't change anything except their marketing messages to include the word "analytics." But even the skeptical view does not refute the fact that there is clearly a lot of demand for analytics—otherwise these firms would not be trying to jump on the bandwagon in the first place.

The term *analytics* has even made it into the popular press. It is common to see it used in the business section and even on the front page of the newspaper. The movie *Moneyball* (based on the book of the same name)—showcasing how the clever use of statistics helped the Oakland A's create a low-cost winning team—might be most responsible for the term *analytics* catching the general public's eye.

Those following politics closely have also likely read about President Obama's re-election campaign's heavy use of analytics to target the voters most likely to vote for him. These forays into mainstream culture (sports and politics) furthered the public's interest in analytics.

Responding to the demands of employers and students' desires to be employable, universities have started to offer degrees in analytics. It is very likely that you have come across articles in major publications that talk about the demand for people with analytics skills.[2]

But just what *is* analytics? Companies and journalists are better at using the word *analytics* than at telling you what it is. Without a definition, how do you know if what you are doing is analytics? How do you apply analytics to your situation? How do you know if your vendors are really even selling you an analytics solution?

Whenever a word has such good connotations yet is poorly defined, it is at risk of becoming just another buzzword that will be forgotten when the next one comes around. The CEO wants more analytics projects, so managers put the word *analytics* into all their project titles, without substantively changing anything they are doing. Vendors realize that they can rename what they've always done as *analytics*, and it will sell better. And if a certain group of products become associated with the word *analytics*, then the companies selling those products have no incentive to clarify; why clarify the term if it is working to your advantage? Part of the appeal may very well lie in the mystery of analytics to some people: It sounds complex and promises great returns, so we better do it!

If you are serious about applying analytics correctly, how do you know where to start? How do you know when and where you are going to get value from analytics? Will any analytics projects do, or are some better than others?

Adding to the confusion, the term *Big Data* is starting to be used alongside *analytics*. The term *Big Data* is also not well defined. For example, how is Big Data different from a large data set? And does analytics only apply to Big Data? Do *analytics* and *Big Data* mean the same thing?

This book cuts through the confusion and gives you clear definitions of *analytics* and *Big Data*. You'll get a look at this newly defined field so you will have a deep understanding of the term.

We take the position that analytics is more than just a buzzword or a fashionable trend. Analytics, performed well by capable people, can bring tremendous value to your company or organization. And it applies to companies and organizations of all shapes and sizes. It applies almost everywhere—to large and small companies, non-profits, and educational institutions. It applies to healthcare and medicine, government agencies, science, law enforcement, and the military. Analytics projects can be started by the head of an organization, a manager of a department, or even a single individual. But analytics can bring value only if you and the people in your organization know what it is, can communicate it clearly, and apply it correctly.

What Is Analytics?

The term *analytics* has been used for a long time. So what has made its use so popular now?

The internet companies of Silicon Valley have helped. They use the term *analytics* to refer to keeping track of who is clicking on your website, which pages they visit, what they buy, and so on. But it is hard to imagine that the term has spread like it has if it's just about tracking website performance.

Some companies selling reporting systems (or business intelligence software, if you want to use the industry term) have also helped. These companies claim that analytics is the ability to report on your data in easier and more powerful ways. But again, the term wouldn't be so popular if it were just about reporting.

We believe that the term entered the mainstream business vernacular when *Harvard Business Review* published the article, "Competing on Analytics," by Thomas Davenport in January 2006 (and then a book by the same title).[3] In this article, Davenport highlights how companies like Amazon, Marriott, Harrah's, and Capital One "have dominated their fields by deploying industrial-strength analytics across a wide variety of activities."

This was a tipping point for the term *analytics*. The article really shows that you can apply analytics to a wide range of problems. And it shows that it isn't just a niche area (like tracking website visitors or creating better reports). It's bigger than that.

But we still haven't actually defined what exactly analytics is. The Davenport article gives examples of firms solving specific problems. For example, Marriott uses analytics to set the optimal price for rooms, and Capital One uses it to analyze experiments with different prices, promotions, and bundled services to attract the right customers. But the article only defines *analytics* as the ability to "collect, analyze, and act on data." In other words, according to this article, analytics is about using data to make better decisions. This definition, while correct and compact, does not give much guidance. Haven't managers always talked about using data to make decisions? They have, but there is now much more data available. Simply using more data to make more decisions doesn't really help us define *analytics*.

The contribution of the Davenport article isn't that it defined analytics. Rather, the article helped create the analytics movement. That is, it introduced the idea that a lot of people are solving a lot of different problems using a lot of different tools—and that all these tools are being called "analytics."

The article sparked the idea of analytics as a field (like the field of computer science or of chemistry). As proof of this, many universities have started to offer degrees in analytics. Professional organizations dedicated to data analysis, like INFORMS, have also started to shape the field.[4]

What has emerged from academic and serious business thinkers is a definition of analytics that categorizes the different objectives when using data to make better decisions.[5] We will stick to this emerging definition of analytics throughout this book. **So, here is the definition of *analytics*:**

> Analytics is the collection of disciplines that use data to gain insight and help make better decisions. It is composed of ***descriptive analytics*** to help describe, report on, and visualize the data; ***predictive analytics*** to help anticipate trends and identify relationships in the data; and ***prescriptive analytics*** to help guide the best decisions with a course of action given the data you have and the trends you expect.

Another way of looking at this is to say descriptive analytics aims to provide an understanding of what happened or is happening, predictive analytics aims to tell you what will or may happen next, and prescriptive analytics aims to tell you what you should do. Each of these areas of analytics can be broken down further—which we do later in the book—and different tools and techniques are applied to each.

This definition of *analytics* will hold up over time. It is specific enough to give meaning to the term, while broad enough to allow for future development. New subcategories are likely to come into existence, new tools will surely be developed, and people will gain new insights. This definition gives you a way to understand those changes and how they fit in.

This book dives into each of these areas of analytics to give you more insight. Keep in mind that although we may cover a certain subcategory of analytics or a single tool in just a couple paragraphs, there could be entire university departments, professional organizations, and companies dedicated to just that subcategory or tool. We are by no means understating the importance of that topic. Rather, our goal is to provide you with enough insight so that you can better understand the field of analytics.

To help understand the types of analytics further, let's explore a few examples.

Examples of Descriptive Analytics

A great early example of descriptive analytics comes from John Snow's work during London's 1854 deadly cholera outbreak. The data on where people were dying was readily available but wasn't helping anyone contain the outbreak. But Snow decided to plot the deaths on a map to see if he could get additional insight (see Figure 1.1). In this map, the small black dots represent the residences where people died from the disease. When Snow and others who lived in London at the time looked at the map, they could see clearly that the deaths were centered around a certain water pump.[6] (As noted in the "Endnotes" section, a modern version of this map, creating with modern mapping tools, shows the data even better.) Looking at the data this way helped narrow the search for the source of the cholera outbreak to a particular water pump.

Of course, in reality, it took quite a long time to convince the skeptics that the water pump was the source. But, in the end, Snow was correct, and his map played an important part in convincing people. Some even say that this was the start of the field of epidemiology (the science of studying the patterns and causes of diseases). In this case, Snow visualized data in a new way, which led to a much greater understanding and helped convince others. This is the power of descriptive analytics.

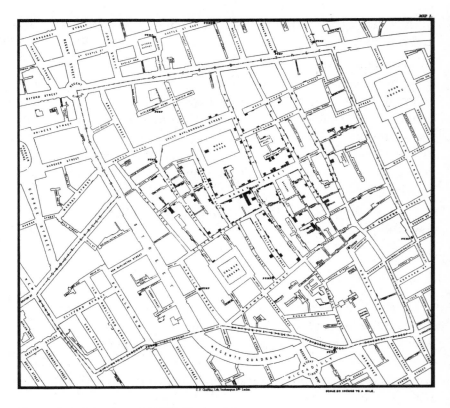

Figure 1.1 John Snow's 1854 Cholera Map

Interestingly, Snow's idea is being applied in a modern way in Lahore, Pakistan, to help prevent the spread of the mosquito-borne disease dengue fever. The city health team is using a smartphone app to record the presence of infected mosquito larvae and plotting this information on a map along with known infections. This information describes the extent or potential for dengue fever, and officials can use it to help determine where to spray. It is a clever update of Snow's idea and shows the power of geographic visualization.

The case of Circle of Moms[7] shows that a little descriptive analytics can lead to big strategic decisions. There was a Facebook application originally called Circle of Friends that had gone viral shortly after it was created and had grown to 10 million people signed up by 2008. Although it looked like they had a potentially popular social website, the founder and CEO realized they had a problem. Almost no one was actually using the application. The founder knew he needed engaged users—not just a lot of people signed up—to have a company with real value. Simple descriptive analytics came to the rescue. The founder began to spend time looking at their data on the users they did have. When he investigated the data, he found that one subgroup was using the application much more than others: moms. The moms were more engaged, had longer conversations, posted pictures, and did other things that indicated they were a happy community group. So the Circle of Friends management team decided to modify the entire business and make it Circle of Moms. Having the ability to look at data and understand it in new ways can lead to big strategic decisions like this one.

Examples of Predictive Analytics

You often interact with predictive analytics through consumer websites, although you may not have known it. Netflix and Amazon are now widely known for their predictive analytics. Based on your ratings (or your purchase and viewing history) and the ratings of others, Netflix and Amazon recommend additional merchandise that you will probably like. It is a bit difficult for an outsider to know how valuable a good prediction system is to these companies. However, we know that Netflix must believe it is pretty important, based on the fact that they conducted an entire contest (available to anyone except Netflix employees or close associates) solely designed to create an improved recommendation algorithm. In 2009, the winning team, which was able to demonstrate a 10% better recommendation system, was awarded the $1 million prize.

Many websites are running tests to predict what designs will work better. In 2008, for example, the Obama campaign was designing its website so that it could maximize the number of people who signed up to be on its email list. Once they were on the email list, people could be contacted for donations or to volunteer. The design team had a choice of several different buttons with different labels (like "Join Us Now," "Learn More," and "Sign Up") and about six different pictures or videos. Instead of arguing about which combination was best, they ran a test. They were getting enough hits to their website that they could randomly assign a visitor one of the different combinations and then track what happened. They ran these tests against a control (the existing design) to predict whether the change would have an impact. In the end, they found a combination that they claimed led to a 40% increase in the number of people who signed up to be on the email list. They claimed that this led to a big increase in donations and volunteer hours.[8]

Examples of Prescriptive Analytics

To borrow an example from the book *The Optimization Edge*,[9] you interact with prescriptive analytics when you use a GPS system or an online map to get directions. You enter the start and end points, and then the program tells you (or *prescribes*) how to get there. Most people don't know that mathematical optimization algorithms are what allow this to happen.

Although you might not think about it, matching kidney donors and recipients is also a prescriptive analytics problem.[10] A person needs only one healthy kidney. Therefore, a healthy donor can donate one healthy kidney to someone who has no functioning kidneys. The alternatives are dialysis and getting a kidney from a deceased person. Waiting for a deceased donor can take a long time, though. Also, the quality and length of life are much better if a recipient receives a kidney from a live donor. Knowing this, a person in need of a kidney can sometimes find a family member or friend who is willing to donate a healthy kidney. However, the problem is that there is a good chance the kidneys won't be compatible; just because the donor and recipient are family members or friends doesn't mean they will be a match. Say that you need a kidney, and your brother is willing to give you one, but you're not a match. If you both get into a database of donor–recipient pairs, you can be matched with a compatible donor, and your brother can be matched with a compatible recipient. You don't actually get your brother's kidney, but he gives up one of his kidneys so you can get one from someone else.

To make a kidney donor–recipient database work, matching organizations use mathematical optimization techniques to prescribe matches. That is, they look at all the possible combinations of matches and pick the best ones that match up the most people and provide the highest possible compatibility matches. The mathematical optimization engine makes this possible. Without it, looking at all the combinations would be impossible.[11] Optimization helps connect many more people and improve many more lives than otherwise would be.

The following examples show companies using all three types of analytics—descriptive, predictive, and prescriptive—to improve their efficiency. As you'll see with both the DC Water and Coca-Cola cases, a lot of the value of analytics results from combining different types of analytics together to come up with a solution that just wasn't possible several years ago.

An Example Using Descriptive, Predictive, and Prescriptive Analytics: DC Water[12]

DC Water, the water company for the Washington, DC, area, serves more than 2 million customers with several thousand miles of pipes and maintains nearly 10,000 fire hydrants. The average age of the pipes is over 75 years. DC Water documented its efforts with IBM to go from a company that used mostly paper records and limited data to an organization that improved performance with descriptive, predictive, and prescriptive analytics. This case highlights how each subcategory of analytics offers value and that they can all complement one another.

By using descriptive analytics, DC Water mapped the location of all the city's fire hydrants. Just visualizing the hydrants allowed the company to create better maintenance plans; before it did, it had been difficult to make sure every hydrant was being properly maintained. DC Water also added extra sensors to water pipes to better monitor water usage and look for anomalies.

With so many aging pipes, failures were a big problem. In the past, DC Water simply reacted to the failures. With predictive analytics, the company could now look at factors such as the age of the pipe, soil conditions (gathered from better descriptive analytics), pressure on the pipes, nearby problems, and other factors and use statistical models to predict when pipes would fail. This allowed DC Water to address potential issues before they became actual problems. It also helped the company prioritize preventive maintenance.

By using prescriptive analytics, DC Water could better route maintenance crews to fix trouble tickets, which increased the productivity of the maintenance team while driving down fuel cost. By knowing the location of each hydrant, existing work order, or preventive maintenance project, DC Water could better route the trucks and crews to the best locations.

An Example Using Descriptive, Predictive, and Prescriptive Analytics: Coca-Cola Orange Juice Plant[13]

Businessweek published an article on Coca-Cola's new state-of-the-art orange juice plant. The goal of the plant was to produce high-quality and consistent orange juice year-round. The article showed how Coca-Cola was using analytics to accomplish this.

As an example of descriptive analytics, Coca-Cola used satellite images of the orange groves in Brazil to determine when different fields were ready for harvest. By being able to see the orange groves in a new way, Coca-Cola was able to better harvest the oranges for increased quality.

Coca-Cola used predictive analytics to predict the quality of oranges coming from different fields by looking at different weather patterns. This helped determine what types of oranges would arrive and whether the company would have to acquire oranges from other locations (if a particular location was likely to have a low-quality harvest). Coca-Cola also analyzed the chemical components of oranges to predict the taste and the range where people could detect a difference. For example, to make up numbers to illustrate a point, Coca Cola would need to determine whether the sweetness range needed to be between 100 and 200 or between 155 and 175 so they could maintain consistency.

Finally, Coca-Cola used prescriptive analytics to determine how to blend the oranges of different quality and characteristics to get the desired output. The mathematical optimization looks at all the potential oranges available and comes up with the mix that gets exactly the right combination of all the different chemical characteristics at the lowest cost.

So, What's New?

Based on our definition of *analytics*, you may be thinking that many of the disciplines within analytics are not new. You are correct. Many of them have been around for years. But many things are new. We'll cover these in more detail as we move through the book. The following list will give you a flavor for where we're going.

First, there is a now a realization that all these disciplines are related and part of the larger theme of analytics. New solutions become possible as you combine different types of analytics to solve a problem.

Second, the proliferation of data has opened the possibility of asking many new questions and uncovering many new insights and trends. Analyzing and summarizing that same data, however, requires sophisticated tools and methodologies because of the data volume and complexity. At the same time, to take full advantage of the opportunity presented, organizations need to be able to deploy tools and methods more widely; in the past, these tools and methods were stuck in the corner of an organization, run by experts. The more people who can look at and analyze the data, the better your chances of finding interesting trends and running your business better. In addition, this proliferation of data has unleashed a new wave of creativity. Early on, website developers realized that they could combine two or more different services or products to create a new one. This was called a "mash-up." More abundant data has made mash-ups accessible to more managers. They can pull data from many different sources to solve new problems or to solve an old problem in a new way. Creating mash-ups requires creativity.

Third, the proliferation of data has given people incentive to create new tools, revisit old tools that didn't work with limited data, and apply old tools in new ways. For example, when you have access to the whole universe of data rather than just a sample of it, you can analyze it with algorithms that might not have worked well with the smaller sample set of data. (We'll give some nice examples of this in Chapter 2, "What is Driving the Analytics Movement?") Also, with large data sets, mathematicians are rediscovering old fields. For example, the field of topology has been around as a purely theoretical field for 250 years. Now topology is being used to help people visualize large data sets. Finally, new tools (or updated versions of them), like machine learning algorithms (which we will cover later), are moving out of research labs and into the hands of businesspeople.

Fourth, with a lot of business moving online combined with the fast feedback of social media, analytics makes running tests much easier. For example, a company can show different versions of its website to randomly selected visitors and easily test which version leads to the desired results—such as more sales, more signups, or longer time on the site. With more business being done online, analytics can help make and influence more and more business decisions.

Another way to look at what is new in the field of analytics is to consider how it can change what managers need to do. The former president of a successful online financial services firm summed it up nicely. He said that his job wasn't to figure out what decisions to make. Instead, it was to figure out how the decisions should be made. Once he was confident in how decisions should be made, algorithms could be programmed to make those decisions. The algorithms ran the business. The algorithms determined what webpages to show each visitor, what services to offer the visitor, and what price to charge. Management simply needed to make sure the algorithms stayed up to date and used additional data as it became available.

Finally, and possibly most importantly, the attitude toward analytics is new: With the abundance of data available and the abundance of tools to use that data, managers realize that more and more decisions can be improved through the use of analytics. If they don't take advantage of that, they risk falling behind.

What Is the Best Type of Analytics?

Often, when people explain the three types of analytics, they present a diagram that shows descriptive, predictive, and prescriptive analytics building on each other. That is, you need to do descriptive analytics first, then predictive, and finally prescriptive. The diagram then usually suggests that descriptive analytics is the easiest but adds the least amount of business value. On the other end of the spectrum, prescriptive is labeled as the hardest but yielding the most potential business value. Figure 1.2 shows an example of this type of diagram.

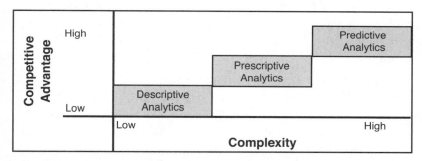

Figure 1.2 Sample of a Typical (but Misleading) Diagram of the Types of Analytics

This diagram is visually interesting and can provoke a good discussion. And we, the authors of this book, used such a diagram extensively in the past. However, we now think that this diagram is misleading. In some cases, the diagram can be true. But, it is not universally true. And it may be true in only a small number of cases.

The value of the different types of analytics is tied to the problems you are solving, not the techniques themselves. Each type of analytics can have relatively minor impact on a business or completely change the business.

For example, if your descriptive analytics project is to just come up with a better understanding of what type of product is sold in what geographic area, you will likely get some interesting insight, but it is unlikely to change your business. If, on the other hand, your descriptive analytics project uncovers deeper insight about your customer, it could cause you to change your entire business, as in the case of

Circle of Moms. This represents a huge strategic change. As another example, good descriptive analytics applied to data sets concerning cancer patients has uncovered potential life-saving treatments for different types of patients. It would be hard to argue that this type of descriptive analytics isn't extremely valuable.

Predictive and prescriptive analytics work the same way. You could have many nice projects that are relatively small in terms of the overall strategy of the firm (but could still be good for that segment of the business) and you can have projects of strategic importance. For example, a small predictive project may involve improving forecast accuracy by 10% for certain items. Certainly this is nice, but it's unlikely to change the business. On the other hand, Amazon's use of predictive analytics to make buying suggestions changed the nature of retailing.

For prescriptive analytics, we could imagine that the routing of trucks for DC Water saved some money but didn't change the business. On the other hand, Coca-Cola's use of prescriptive optimization to blend orange juice to improve the quality of the product could have significant strategic impact on the product. In another example, Indeval, Mexico's central securities depository, uses prescriptive analytics to help settle securities. The system determines how to match buyers and sellers and clears $250 billion per day, saved $240 million in interest over 18 months, and made the market much more liquid. Indeval says that the system was in place during the stock market crash in 2008, and the extra liquidity in the system allowed people to exchange securities more often during a single day—which was much better than in other countries, like the United States, where the lack of liquidity meant people were stuck holding securities for an extra few hours as the price of the security rapidly decreased. Indeval's use of prescriptive analytics is obviously very strategic.

It is also impossible to say which type of analytics is easiest to implement. You can have a simple descriptive analytics project where you load the data you have into a better reporting tool and immediately gain insight. Or you could spend two years implementing a full-blown descriptive analytics system that gives you access to every bit of data in your organization. Likewise for predictive and prescriptive analytics: You can do good work in an Excel spreadsheet or you can custom build systems that require years of effort and huge teams of people.

Finally, there is no particular order in which to implement these systems. You might think you need to do descriptive analytics to understand what is going on in the business first. But if you have data (which most firms do) and know what problem you want to address, you can skip descriptive analytics and start directly with a predictive or prescriptive project. And, in practice, this is often what happens. The types of analytics projects you do depend on the business issues you need to address, the value of the project, the ease of doing the project, the skills of your team, and many other factors.

We wish there were a simple roadmap that you could follow to get the most out of analytics. But the wide range of analytics applications and tools is what makes the field so fascinating and rich. You have many different options and have to pick the best approach for your business for each project (big or small).

What Is Managerial Analytics?

We use the term *managerial analytics* as the title of this book, so we should define what it means. We define *managerial analytics* as what a manager needs to know about the field of analytics to make better decisions. So we are using the term to refer to understanding the field of analytics from a manager's perspective. Or, in other words, *managerial analytics* helps prevent you from being fooled or confused by the many different analytic terms and solutions offered, helps you see what is possible, and helps you do more on your own.

Managerial analytics is about understanding the difference between descriptive, predictive, and prescriptive analytics. It is about understanding the different tools and where they fit in. It is about cutting through the buzzwords to help you better understand solutions.

Hopefully you are now sensing how large the field of analytics really is. You can get an advanced degree in very specific niches within the analytics field. As a manager, you cannot understand every nuance. You should, though, understand what problems the different disciplines of analytics can solve. And you should know what types of decisions a particular project will help you with.

Another way to think about this is that you will often be presented a single tool as a way to do analytics. We want this book to help you, as a manager, cut through the confusion and determine which type of analytics you need and for which types of projects a tool may provide value. No one solution will cover the full range of the field of analytics. This knowledge should help you select analytics projects that meet your objectives and then guide successful implementations.

Managerial analytics is not devoid of technical material. As the importance of data and analytics increases, you will have to be more comfortable with the technical aspects if you want to succeed as a manager. While you won't be expected to understand all the nuances of different disciplines, you should understand the limitations of a given analytics solution—and what that solution won't do. Many times, an analytics solution will be presented by someone pushing an idea, and it will be impossible to tell what the solution won't do. Vendors commonly write descriptions of their analytics solutions that make it seem like a single solution will solve all your problems. If only it were this easy.

This book covers enough of the key technologies behind descriptive, predictive, and prescriptive analytics so that you will know if a project is on track, you will know what experts you may need, and you can better understand the details of the solutions being presented to you and have a healthy understanding of their limitations. In other words, with this understanding, you are less likely to be fooled or confused by vendors and co-workers using the term *analytics* in a vague way.

But managerial analytics is about more than not getting fooled. It is also about the art of the possible. The analytics movement is real because it produces real value. To capture this value, however, you need to know what is possible. So managerial analytics is about understanding how analytics can apply to your business. One of our goals with this book is to get you thinking about your company or organization in a new way. That is, by seeing different examples and understanding the different areas of analytics, you may uncover opportunities for adding value in ways you hadn't thought of before— and these new ideas may turn out to be much more important than current projects you are working on. If you want your entire company or just your small department to make better use of analytics, the more you and your colleagues understand what is possible, the more value you will find.

Finally, not everyone has access to a large IT department and teams of analytics experts. In fact, most people don't. So don't think of managerial analytics as just something that big firms do. Instead, managerial analytics is for organizations of all sizes. Managerial analytics is about giving you enough information that you can get started with what you have. From what we have seen, most managers have plenty of room to use the data and tools they have in order to make better decisions. Don't wait for big projects; you can start now.

The rest of this book is devoted to helping you understand managerial analytics—analytics from a manager's perspective. This will help you whether you are doing a single project within an organization or rolling out analytics solutions to your entire organization.

It will help you understand the definition, help show you how the types of analytics work together, and present you with practical applications that are applicable to a wide range of organizations. It provides cases and examples to solidify the ideas throughout. And, since analytics is a technical field, this book also discusses the technical aspects of analytics. You can skip these sections without losing the flow of the book. But, as a manager, the more you understand the technical aspects, the better you will be able to apply analytics.

Is Competing on Analytics a Strategy?

Davenport's article, "Competing on Analytics," was not written to define analytics. It was written to show that firms that are dedicated to using analytics to tackle the biggest issues and using analytics throughout the firm could do significantly better than other firms in their industry. In the article, he also discusses ways a firm can structure itself to be an analytics competitor. It becomes clear from the article that analytics could be a highly strategic tool for any firm.[14]

An article in *Analytics Magazine* stated, "In many ways, business analytics is the next competitive breakthrough following business automation but with the goal of making better business decisions, rather than simply automating standardized processes."[15]

And you can find many other similar claims.

Based on all this, it might be easy to declare that competing on analytics can be classified as a strategy. A CEO may try to claim "We will compete by being the best at analytics."

Ben Reizenstein, from Northwestern University's Kellogg Business School, actually posed this exact question: Is analytics a strategy? Or, is it just a tool to help you do what you do, but better? He was leaning toward the latter. That is, with all the hype around analytics (much of it deserved), we might have lost track of the idea that analytics can help support your strategy. So, maybe it is not a strategy but a great tool to help you execute your strategy.

For example, "Competing on Analytics" uses Marriott as an example. For a hotel chain, the strategy might be to provide the best business hotels or the best resort hotels, or to be in every market or to be only in large urban business districts. The use of analytics is a way to help the hotel chain really execute its strategy, but its commitment to analytics is not necessarily the strategy itself.

If analytics is a strategy, then you have to wonder if every analytics project has value. If analytics supports your strategy, you can then judge analytics projects as they relate to executing your strategy.

In Davenport and Harris's book, *Competing on Analytics* (the follow-up to the article), they mention that analytics is not a strategy but that it supports developing certain business capabilities.[16]

On the other hand, consider again the hotel example. You could imagine a scenario where a hotel decides to be the best business hotel and will use analytics to do just that. This could be contrasted with a strategy of being the best business hotel by focusing solely on hiring service-oriented people or having a unique sales force to win corporate clients, or to focus on buying distressed hotels in good markets to keep real estate costs low. The list could go on. There is nothing that dictates that a firm needs to use analytics to be better. In this sense, analytics is an integral part of the strategy.

It dictates the type of people the firm hires and the investments the form makes in IT, for example.

Davenport and Harris mention that the companies that are best at analytics mention it in their annual reports and press releases, and it is highly visible to the top of the organization. They even mention that the firm's strategies are built around analytics—suggesting something possibly a bit closer to a strategy than just a supporting role.[17]

Another way to look at this issue is to divide strategy into the business strategy (what the business is trying to be) versus operational strategy (how to achieve the business strategy). With this view, clearly analytics is an operational strategy that supports the business strategy.

We want to present you with a balanced view that neither over- nor undersells analytics. Our aim with this discussion on strategy is not to provide an answer but hopefully to stir a lively debate in your organization about the role that analytics will play.

2

What Is Driving the Analytics Movement?

Data Is the Fuel for Analytics

It is safe to say that without the recent dramatic increase in the amount of data available, analytics would not be such a popular field. Data is the fuel for the analytics movement.

In fairness, even before this recent influx of data, firms were already collecting vast amounts of data. Credit card companies were tracking every transaction by each of their cardholders; cell phone companies were tracking every call; Google was tracking every search; Walmart and Amazon were tracking every purchase; and hotels, casinos, and grocery stores used loyalty cards in order to track their customers' preferences. So it is no surprise that those firms were already engaged in many analytics projects well before the term recently became so popular. And, of course, due to the popularity of those firms and publications about their successes, others began quietly collecting and using more data in their businesses as well.

But several events in the closing years of the 1990s and opening years of the 2000s led to the dramatic increase in data that we are now seeing.

Prior to January 1, 2000, many large firms upgraded their IT systems to avoid the Y2K bug. These systems are typically called Enterprise Resource Planning (ERP) solutions, and they are sold by the likes of SAP, Oracle, Infor, and many others. An ERP system contains all of an organization's financial data, sales data, production and procurement data, human resource data, and so on. In its ideal implementation, an ERP system has all the data you need to run your business in a single place. By the mid-2000s, these systems were in place and collecting massive amounts of data.

In addition, during the first decade of the 2000s, sensors that were connected to the internet became more prevalent. We already mentioned DC Water adding sensors to its network of pipes. Since the beginning of the 21st century, companies have been adding sensors to machinery and industrial equipment. GE recently announced that its new jet engine will create a terabyte of data on a single cross-country flight and the company is therefore interested in investing in analytics. Jones Lang LaSalle, a large commercial real-estate firm that manages many buildings, has developed a solution to monitor all the equipment in a building (like heaters, air conditioners, fans, and so on) to help reduce energy and maintenance costs. Dairy farmers are putting sensors on their cows to monitor their health. Although this may sound strange, a cow is a large investment, and keeping tabs on its vital signs can help a farmer keep the cow healthier and more productive. RFID tags are placed on retail products to allow them to be tracked, in real time, from the supplier to the store shelf. It's safe to say that we haven't seen the end of what people will attach sensors to.

Online transactions also began to skyrocket in the early 2000s. Managers of websites realized that they could track customer click patterns, how long customers looked at a new product, and how long they stayed on the website. This allowed firms to customize the shopping experience to each shopper and show additional items a particular person might be interested in.

Smart phones became another source of data. They have the ability to provide information on location and identify you while you're in a store. For brick-and-mortar retailers, headaches ensued when customers in the store would simultaneously research online options with a competitor. For example, Best Buy was worried that it was becoming simply a showroom for Amazon.com—that is, buyers would evaluate an item at Best Buy and then buy it from Amazon.com.

Social media sites also started to take off in the early 2000s, creating a stream of data directly from consumers. This data, however, was different in nature from much of the traditional data that companies would store in that it wasn't neatly structured. It consisted of pictures, videos, and text (with lots of typos, misspellings, and questionable grammar).

So managers realized that they had access to more data than they had ever had access to before. Even firms that were collecting large amounts of data in the 1980s and early 1990s realized that the new sources of data would be valuable to them. Naturally, managers were left wanting for real value from the data they now had access to.

In fact, the value of this data led to claims that data was perhaps just as valuable to a firm as labor and capital. That is, data could be thought of as another key economic input.[1] So it is no surprise that as the leading firms started to extract value from their data, other firms quickly followed suit.

It is difficult to pinpoint an exact time, but around 2005, the timing was ripe for a data tipping point. That is, around 2005, the amount of new data began to skyrocket at a rate that became impossible to ignore. By 2008, as evidence of the explosion of data, *Wired* magazine had a provocative cover story titled "The End of Science," which argued that the abundance of data made the scientific method obsolete.[2] The scientific method relies on formulating a hypothesis and then gathering data to test the hypothesis; if the hypothesis stands up to the testing, it becomes a theory. Now, the *Wired* article argued, you could simply analyze these vast data sets and draw conclusions. These conclusions would become the new theory because the data sets presumably included "all" the information.

It was this explosion of available data that helped create the need for analytics.

What Is Big Data?

The increase in the amount of data available has captured the public's imagination. The term *Big Data* is frequently used in the news. Although the origin of the term *Big Data* is debatable, it was likely used by technical people to describe the fact that there was something fundamentally different about these new data sets. The term has escaped into the popular lexicon. We suspect the reason is that it is so similar to George Orwell's "Big Brother." Recent news stories about government programs to collect and analyze this data have increased general awareness. People are becoming more aware that lots of data is being collected about them—in some cases very willingly but in other cases quite without their consent.

It is also likely that the term *Big Data* has stuck because it correctly describes some fundamental shift. With Big Data, you have crossed some threshold where traditional data storage and retrieval no longer work. In other words, you have so much data that it's broken the databases. So the term is useful for IT professionals to describe different types of data.

Google, Amazon, Walmart, and other large firms (like banks and phone companies) were probably the first to encounter problems with Big Data breaking their databases. Large firms had always dealt with large data sets and managed their IT departments accordingly. For example, large banks, the largest retailers, and large internet companies, such as Google and Amazon, had large IT departments and spent a lot of money on computer equipment and software. But none of that captured the public's imagination, and no one really heard much about it.

But soon smaller firms and many other types of organizations realized that they also had problems with the proliferation of new data, and the term *Big Data* caught on. And it wasn't all just about the size of the data. As with the term *analytics*, the definition of *Big Data* can be difficult to pin down.

We have found three good definitions of *Big Data* floating around but no consensus about any of them. Some people may even use the term *Big Data* to refer to what we are calling *analytics*. We want to be sure to keep the terms separate in this book.

Here we will spell out the three definitions of Big Data so you can have a framework for understanding what someone might be talking about if they mention Big Data.

The First Definition of Big Data

First, there is an IT-centric definition of *Big Data*. This definition sticks close to the term's roots. It simply means that there is a lot of data—in fact, there is so much data that it breaks traditional data storage and analysis technologies. This obviously plays off the term *big*. In his book, *Taming the Big Data Tidal Wave*, Bill Franks intentionally avoids providing a firm definition and stuck with the fact that Big Data is just bigger than current technology can handle.[3] He realizes that today's Big Data sets will eventually find technologies that work, but those technologies will again not suffice with the next wave of data.

Gartner, IBM, and others take this same definition a little further and classify Big Data using three Vs. They think that Big Data can break existing technologies in different ways, and they want to capture that nuance with the three Vs:

- **Volume**—This is the V we think of the most: just having so much data that it requires a different technology. In other words, you have so much data that you fill up your existing servers, and your existing databases break when you try to use them.

- **Velocity**—This category refers to the fact that data is arriving very quickly. Often, it also needs to be analyzed as fast as it is received. This data mostly comes from sensors and other real-time devices. With some traditional large data sets like credit card history, you could store the data and then summarize and retrieve it at a later time. But with sensors feeding streams of data in real time, you may need other technologies for monitoring this data so you can take immediate action. For example, if you are monitoring 10,000 sensors in a chemical plant, and these sensors are feeding you multiple bits of information in intervals of less than a second and are looking for signs of quality or machine failures, you need a way to handle all this.

- **Variety**—This V stems from the fact that data is coming from many sources and is in many different forms, like text, videos, and pictures. And a lot of the data is not structured. For example, at IBM, the creators of the *Jeopardy!*-playing Watson computer realized that there was a vast amount of data on cancer treatments in research papers and medical reports. Although this data could not be stored or retrieved in a traditional manner, you can imagine the value it could provide. The IBM folks realized that if someone could get to this data, they could ultimately help doctors understand cancer treatment in a much deeper way.

Occasionally you will also see companies touting other Vs, such as veracity, meaning that you need ways to validate the data. This is often the case when data comes from outside a firm. It is very likely that others will expand on this list in important ways, but in any case, the idea is that this definition of Big Data refers to just having a lot of data.

The Second Definition of Big Data

We think the second definition of Big Data is the most thought provoking. It comes from Victor Mayer-Schonberger and Kenneth Cukier's book, *Big Data*. In this definition, *Big Data* means you've captured the entire universe of data for the problem you are looking at.

Mayer-Schonberger and Cukier correctly point out that much of what was previously developed in the field of statistics was meant to draw good insights from small random samples of data. So, what they are truly claiming is that we are entering an era where we no longer have to sample the data; instead, we can analyze the full universe of data. This definition opens up some interesting insights.

First, these authors claim that algorithms may be much more accurate when they are working with the universe of data than when they're working with a sample. One example involves the work on creating good algorithms for translating text from one language to another. Despite decades of work, not much progress was made. The researchers at Google and Microsoft realized that if they fed the same algorithms much more data, the algorithms were much more accurate. That is, people who spoke both languages started to agree that translation was good. So, instead of having a million reference translations, they could do much better with a billion, and they could do significantly better with a trillion. This extra data allowed the algorithms to pick up more on all the nuances of the different languages.

Second, having a full set of data may also allow you to detect patterns that you couldn't see before. For example, the book mentions a project in Canada that stored and analyzed premature babies' vital signs. Previously the project had just analyzed this information at the time of the readings but had not stored the data—which is technically a lot like sampling. When the project participants analyzed all the data, an interesting pattern emerged. They noticed that several hours prior to an infection, all of a baby's vital signs were unusually normal—that is, all vital signs were perfectly normal, which is unusual for a premature baby. They could not see this pattern when just sampling the data. In the past, the doctors and nurses may have seen normal vital signs and felt fine and attended to other issues. But now the medical staff was on alert and could start working on the baby to minimize the impact of the infection.

Another way to say this is that when you have a full set of data, your concern lies only in the interpretation and analysis of the data rather than also needing worry about whether the sample of data is sufficiently large, unbiased, and representative of what you are trying to study. Also, it is through cases like the Canadian example that people make the claim that you only need to worry about correlation and not causation. To return to our example, it doesn't matter why a baby's vital signs look unusually normal prior to infection. It just matters that it happens. And knowing it allows the medical staff to take action. Of course, natural curiosity leads some people to want to find out the *why* behind such a finding. But that, the book argues, is really of secondary importance.

The book *Big Data* occasionally seems to dismiss the importance of statistics. We think this is a mistake. Statistics is a well-developed and practical field and is an important part analytics. You may use less of traditional statistics when you have a full universe of data, but traditional statistics certainly won't go away. But the book does make an important point that doing statistics with samples works very well with random samples—and *random* is the key word here. Mayer-Schonberger and Cukier argue that in reality, it is very hard to get a good random sample without bias. The book points out that if you can collect and work with the universe of data, you remove the problems with finding a random sample.

Third, if you truly believe that data has economic value, then owning the full universe of data is a nice position to be in. In fact, in some cases, the secondary act of collecting data may turn out to be more important than the primary business. In *Big Data*, the authors give the example that a lot of what Google does helps the company own a larger data set that it can then leverage to create more value.

Also, in this definition, the Big Data set does not have to be large, although it usually is. For example, you may be studying some particular issue, and the full universe of data related to that might happen to fit within a spreadsheet. So the size of the data is not that large, but because it is the universe of data for that issue, it could still be labeled as Big Data.

We realize that there are some problems with the concept of the universe of data as Big Data. For example, it seems impossible to believe that you could collect data about everything that could ever happen for a given issue. So even though you may be collecting a lot of data, that data could still be biased or incomplete in some way. But we like this definition because it opens up new ways to think about the data you should be collecting.

The Third Definition of Big Data

A third definition of Big Data comes from the popular press. Whereas the first two definitions are concrete and define Big Data as something different than normal data sets, this third definition is turning the term *Big Data* into a buzzword. When you look at how the popular press uses the term *Big Data*, you see that it does not fit the first definition of overwhelmingly large data sets, and it does not fit the second definition of the universe of data. However, we shouldn't dismiss this third definition too quickly; these articles are usually quite interesting and usually point out some new insight. Therefore, the third definition of Big Data is the creative use of new data sets.

Although this definition of Big Data is not as concrete as the other two, we feel that it is an insight worth keeping. Collecting data is becoming cheaper all the time, and standard economics predicts that when something becomes cheaper, we demand more of it. The key for managers is to discover new sets of data that can provide value to an organization. In addition, it may be possible to bring two unrelated data sets together for new value. Rather than being hung up on a precise definition of Big Data, it may be better to think of Big Data as a challenge and an opportunity to leverage data that is available in new and creative ways to make better decisions and add value to your organization.

Can Big Data Replace Science?

At the start of this chapter, we mentioned a *Wired* magazine article that argues that Big Data is trumping the scientific method. The idea is that the large data sets actually suggest what the theory should be. The book *Big Data* suggests that only correlation matters, not causation. The argument is that if you have the universe of data, if you find correlations, you don't need to worry about why the data is correlated. (In our example, the doctors don't need to know why a certain set of vital sign readings means an impending infection; they just need to know that an infection is coming.)

These are provocative ideas. However, we would rather proceed with caution. Finding that large data sets are useful doesn't mean that the scientific method is invalid.

Justin Fox wrote a blog post at *Harvard Business Review* that countered the claims that large data sets are trumping science.[4] His argument is that the scientific method, described simply, is based on testing hypotheses that can be falsified. He claims that large data sets have not made this approach obsolete. Instead, he argues that claims made with large data sets or about large data sets trumping science often have many hidden hypotheses. For example, a hidden hypothesis could be that the patterns we see in a large data set will always hold in the future or that the large data set is in fact the universe of data. Ignoring these hypotheses certainly doesn't make them go away; it may, in fact, lead to bad decisions and doesn't naturally lead to finding a new theory.

Nassim Taleb devoted his book *Black Swan* to the fact that it is impossible to forecast rare events. The idea is that if all you have ever seen is white swans, then it is impossible to predict the existence of a black swan. He also claimed that models built on all the data you've ever seen would still not be accurate because surely there would eventually be a set of events that would lead to a future outlier data point that would nullify your model. Taleb's book was focused on financial modeling, and he put his ideas into practice. By going against traditional financial models, which assume that the future will be like the past, he was able to make a lot of money by realizing that outlier events will occur.

So, our caution with Big Data (as defined as a lot of data or the universe of data) is that it may deliver many valuable insights. But it will not replace thinking.

Can You Do Analytics Without Big Data?

It is important to remember that you can do good analytics without using Big Data. Your data set doesn't have to be large, doesn't have to be the universe, and doesn't even have to be that creative.

In his book, *Taming the Big Data Tidal Wave*, Franks asks what is the more important component: big or data? His answer is neither. He says it is the value you get out of analyzing and taking action on the data.

Descriptive, predictive, and prescriptive analytics can all work on any set of data. Our definition of analytics does not worry about whether the data you are using is "Big Data" or just a normal data set. You might need to use different techniques, depending on the size of the data set, but the type of analytics will still be descriptive, predictive, or prescriptive. For example, we have seen examples of companies doing sophisticated production planning in Excel. On the other hand, Walmart enters 1 million customer transactions into a database sized at over 2.5 petabytes.[5] This large data set cannot be analyzed with Excel.

We have seen that many organizations are sitting on valuable data sets that they are currently not using. You may find that you have data sets you can immediately start analyzing. And this step may help you better define what new data you would like to eventually collect and analyze in the future.

Testing Helps Ignite Analytics

As we saw in the previous section, the increase in the amount of data helped fuel the interest in the field of analytics. But it wouldn't be fair to give data all the credit for the interest in the field of analytics. Some of the credit should go to our belief in the scientific method and its reliance on creating testable hypotheses.

Science advanced rapidly because of the simple idea that we should develop testable hypotheses and then validate each test with data. With the success in science, the idea has crept into business and organizations as well. The implementation in business isn't always as rigorous as in the sciences, but the idea has long existed in business that we should test our ideas with actual data. So, as data became easier to collect, it found a receptive audience from business leaders who wanted to use that data to test more ideas. That is, business leaders knew that they could make better decisions by relying on the results of tests, backed up with data, rather than on their intuition.

Testing naturally took off with the rise of internet companies. Companies quickly realized that they could run tests at almost no cost. They could test whether a green button or an orange button would lead to more clicks. They could test the impact of a change in price on the sales of an item. And, if they had enough traffic to the site, they could get feedback almost immediately. We'll discuss more of this in Chapter 6, "Predictive Analytics," in the section on A/B testing.

For example, Airbnb (pronounced "Air B&B," after a traditional bed and breakfast) is an online firm that matches independent people who want to rent a room or house with buyers. The company ran tests with professional photographs on its website.[6] It thought professional photographs would help give buyers more confidence. So Airbnb set up a simple test to take some professional pictures of some rooms and compare them to a control group. The statistics predicted that this would lead to a two- to three-fold increase in bookings. Airbnb could then more confidently spend extra money to take more pictures, having good confidence in the payout. It was a good predictive test that allowed the company to determine whether it would work. So, instead of arguing about whether professional pictures would work, Airbnb did a scientific test to try the idea.

Keep in mind that using the scientific method in the business world is not new. Using the scientific method to test a hypothesis is a powerful idea that has been on the minds of executives for a long time. Jim Manzi does a nice job of the laying the foundations for testing and gives more examples of more traditional business tests in his book, *Uncontrolled*.[7]

Manzi's book points out that the medical field has long applied the ideas of the scientific method in the form of randomized field trials. For example, to determine whether a new drug works, you carefully select a random group of patients and give half the group the treatment and the other half a placebo. Then you see if the treatment works. The key idea is that you want a random group. The chemistry of biology is complex, and by selecting a random group, you are effectively holding everything else constant and testing the treatment. If the group wasn't randomly selected, there could be some underlying (and unexplained) reason for some of the patients to respond better than others.

Manzi argues that randomized field tests should also be applied to business decisions. The business world is just as complex as biology. A randomized field test is a great way to hold everything else constant while testing a single idea. Of course, it is harder to do this with physical tests (for example, changing the layout of a store) than with internet tests (for example, changing the color of a button). Even though physical testing is more expensive, many companies do, in fact, physically test their ideas. (They were certainty influenced by the scientific method.) If you want to make your own such tests more rigorous and scientifically meaningful, consider the method for randomized field tests.

With a lot more data available, managers are demanding that more ideas be tested. The field of analytics provides these tests. So, in a sense, testing and the scientific method should also get some credit for igniting the interest in the field of analytics.

3

The Analytics Mindset

Successful analytics projects don't just happen automatically. They have to run and used by people within an organization. The people running analytics projects have to carefully select the right tools, carefully set up the models, and do the analysis so that others can understand and trust the results in order to take action.

In other words, to use analytics efficiently, an organization needs to be able to analyze problems in a logical way, think clearly about the data, and have the right mindset. This is easier said than done. And, because it important to get it right, this chapter is devoted to helping you and your organization develop the right mindset for any analytics project.

Managerial Innumeracy

To help you understand what it means to be a company that uses analytics effectively, it might be helpful to understand the opposite: managerial innumeracy. If analytics requires the ability to use and think about numbers in a logical way, managerial innumeracy is the inability to use and think about numbers in a logical way.

We are borrowing this concept from John Allen Paulos's great book called *Innumeracy*. He discussed mathematical illiteracy and resulting bad decisions that people make because of it. Interestingly, he reports on a trend still seen today: highly educated people bragging or talking about their mathematical illiteracy without embarrassment. For example, it is common to hear someone say, "I'm not a numbers person." It is hard to imagine an educated person bragging that they are not a "word" person (and can't read). Paulos makes the point that if you are innumerate, it is easy to get fooled by advertisements, politicians, or your own finances.

Innumeracy inside a business or an organization can be subtle because quarterly or annual reports must be rigorously completed. At the very least, a few people have to understand these numbers. But if a significant number of people in the general population are innumerate in their personal lives, they are likely to bring that to their organizations as well.

Innumeracy inside an organization shows up in a variety of ways. It is most obvious in organizations where the only people who truly understand numbers are a select few accountants who prepare the financial reports. It can also show up when individual managers are innumerate and simply do not feel the need to make decisions supported by data and analysis. But more subtly and maybe more dangerously, because no one realizes it, innumeracy shows up in organizations where people are apparently dealing with numbers and data on a regular basis and yet do not fully understand those numbers, do not sense when the numbers are not internally consistent, and therefore do not see that the numbers are used incorrectly.

The next section digs into a few examples of innumeracy.

The Illusion of Numeracy

Some organizations can fool themselves with what we call the "illusion of numeracy." This illusion presents itself when a firm seems to use numbers to make decisions but does not question the calculations or assumptions that went into the final numbers. For example, one of the authors did an internship with a manufacturing company that was required to show a 20% return on investment before buying any new equipment. The manager wanted a new machine and assumed that it would be 25% more efficient than the machine it would replace. When the manager used the 25% increase in efficiency, the return on investment was only 15% for the new machine. The manager's simple answer to this was to rerun the return on investment calculation but change the increase in efficiency to 40%. That's what he did, and the purchase of the machine was justified. On the surface, this firm used numbers to make decisions, but clearly it didn't use a rigorous or very trustworthy process. In this case, it is clear that this was an illusion: Numbers were simply fudged in order to get the result that agreed with the manager's preconceived decision.

Another example of the illusion of numeracy is the use of numbers to score different alternatives, weight the numbers, and come up with an answer. This is how business schools are ranked in magazines. Several different factors are considered (starting salaries, publications by the professors, professors per student, and so on). The magazine then scores each factor, adds them all up, and presents the list. It looks scientific, but there are two major problems. First, it is often the case that several categories are scored subjectively. Second, and perhaps more importantly, the weights given to each category are arbitrary. Applying different weights to the factors could produce a completely different ranking. It is, in essence, an illusion of numeracy. Now, if the publication listed the top business schools by one factor, that would clearly be legitimate, but these rankings strive to provide an overall ranking. Companies also tend to do this when making investment decisions with multiple choices. They commonly either score different factors for each choice and add them up or simply draw up a list of "pros and cons" and debate that list. In the last case, each person still has a relative weight they are assigning; it is just rarely articulated. You may argue that some of these decisions are tough—and you are correct. But you can use techniques like the efficient frontier, which we'll cover later, to plot the outcomes with different weights for the various factors and then make more educated decisions. You still will have tough trade-offs, but at least you aren't fooling yourself with a single numeric ranking that is meaningless.

A subtler version of the illusion of numeracy is seen in the popular quote "Do you think it, or do you *know* it?" This quote has been used to show that a manager is driven by data and analytics—the manager wants more than just an opinion. In this way, the quote can serve a good purpose. But often we see the quote used to demand that someone *know* the answer with certainty (often to hold someone accountable for the number). This is an incorrect read of the quote and leads you down the path of the illusion of innumeracy. You may know that you shipped 100 units last month. And you may have done extensive analyses and determined that you are likely to ship 125 next month. Unfortunately, there is no way for you to *know* that with certainty. If you've done good predictive analysis, however, you should be able to say how much confidence you have in a forecast and what the potential range could be. Although your confidence is not a fact, it is still valuable information when making a decision. We shouldn't force ourselves to pretend to know things that cannot be known. Instead, good analytics demands that we talk about the confidence in an answer and the range of likely outcomes.

The illusion of numeracy can also be seen in sales forecasting for businesses that sell high-priced, low-volume items—such as an airplane manufacturer that sells just a handful of units in a period or an industrial construction company that wins a few new contracts per period. Some organizations will forecast future performance in a quarter based on the performance from the same quarter in the previous year's results. This number then becomes the budget, or "the number," for the quarter. Since one or two big transactions can sway historical numbers significantly, these numbers may not be very accurate. Although these forecast numbers can be used for strictly motivational purposes, managers should always be aware of the accuracy of these numbers and not let them slip into other budget planning activities.

The illusion of numeracy also appears in commodities markets that experience wide swings in prices and demand. Jack Welch, the former CEO of GE and now frequent writer, told a story about how predictions in a commodities market could disrupt planning activities.[1] As the story goes, Mr. Welch was mostly managing very stable industrial business units. He would give each business unit some target—like improve the profit margin by 1% for the quarter. Each quarter, each business-unit managers would work hard to deliver on the 1% and report on the results. Some would come in at 0.7%, some at 1.4%, and so on. This would translate into several million dollars more in profit, and the process would start over for the next quarter, with another target of a 1% improvement. This was an effective way to make the business better and forecast future results. Sometime later, GE bought a commodities company that was in a market that experiences wide price swings. When it was this new manager's turn to report the quarterly improvements, it was useless for him to use the 1% improvement as a goal. Based on the commodity prices in his market, he may have lost $300 million or made $500 million in profit.

In other words, forecasting of steady improvements based on the last applicable period did not apply to this business. Mr. Welch admitted that maybe this wasn't a business that fit easily with GE's portfolio: It needed a different forecasting and planning method. To forecast it like the other business units would have just been an illusion of numeracy.

The Filtration Fallacy

When working with sophisticated analytical techniques, it is important not to overestimate the level of accuracy you are applying. Certain systems can be studied within a high precision, and for them, very fine-grained numeric resolution is not only possible but, in fact, required.

Financial systems provide an obvious example here. Think about when people talk about the "estimation" of Warren Buffett's net worth. Most people are referring to the limits of publically available knowledge rather than to the limits of human knowledge. Whereas a reporter for *Forbes* magazine can only guess at the extent of Mr. Buffett's money market position, you can rest assured that Mr. Buffett (or, more likely, his accountant) knows the exact amount to the penny.

Similarly, this area of finance leads to stories, be they true or apocryphal, regarding fine-grained numerics resulting in significant yield. In popular culture, the idea of squirreling away a fortune composed of ill-gotten "fractional pennies" is a common trope of undetectable white-collar crime (see the movie *Office Space* for an example). High-frequency commodity trading involves very narrow margins adding up to a notable revenue stream. In general, transactional systems can often be understood with such high precision that any improvement, however minor, is worth pursuing.

However, such lessons are often generalized well beyond their appropriate domain. In prescriptive analytics, this idea can be expressed through a misconception we call the "filtration fallacy." Such a mistake consists of believing that a very minor improvement can always be "filtered" through to the bottom-line performance.

Consider a company that enjoys a profit margin of roughly 5%. That is to say, its annual budget typically involves spending $95 million in expenses and receiving $100 million in revenue. A careful analyst might propose changing some aspect of the business that is predicted to reduce logistics cost by $2 million. The supply chain manager might balk, saying, "$2 million is a lot of money to you and me, but it represents only 2.1% of our typical expenses." The analyst responds by saying, "Sure, that's 2.1% of your budget, but those savings go right to your bottom line. My change will increase your profits by 40%."

In this case, the analyst *might* be applying the filtration fallacy. He is assuming that $2 million in projected savings can be filtered directly into the annual profit calculations. For some changes, this is a reasonable assumption, but for others it may reflect a naïve understanding of numeric precision.

Suppose that this $2 million in savings is achieved by applying some mature, well-understood technological improvement. Perhaps it is derived from improving the mileage of a truck fleet by applying sensors to ensure that the tires are always properly inflated. The analyst might be basing the savings numbers on the average annual miles traveled by the truck fleet in question and on typical efficiency improvements seen when rolling out the sensors. In this case, a savings of roughly $2 million might be quite a reliable bet. As long as the trucks experience the typical mileage improvements and travel the typical total annual distance, the profit yield will be substantial.

However, now imagine that this savings is coming from a much more speculative change. Suppose the analyst is proposing a complete reorganization of the structure of the "hubs" of the hub-and-spoke truck routes. (Like the airline industry, the trucking industry uses hubs to move product between major cities and spoke facilities to handle local routing.) The new routing scheme will change dramatically which long-distance carriers are used to bring goods into hubs and which routes carry the goods outbound on trucks. Assuming that everything works as planned, this new system will improve total expenses to the tune of $2 million and profits by 40%. In this case, there is clearly a greater opportunity for things to go off script. Perhaps customer demand will deviate from expectations, forcing shipments into areas poorly serviced by the new hubs and away from areas best positioned to exploit them. This company has little experience using the inexpensive bulk carriers inbound to the new hubs. Their reliability might fall short or their prices might go over their bids. The new hubs may also be seen as undesirable work locations and spur a few key employees to find work elsewhere. Any (or all) of these hiccups could easily wipe out the $2 million in predicted savings, causing profits to plummet rather than skyrocket.

Of course, every change in life is full of unexpected consequences, but some changes clearly court the unexpected more than others. The filtration fallacy assumes that improvements will filter through the system, unmolested by the negative side effects of unforeseen events. The fallacy here is more philosophical than mathematical, however. When we commit this fallacy, we are assuming that we understand our system enough to reliably execute even small improvements. When these improvements themselves are facilitated by small changes, our assumption is quite reasonable. When the agent of change is widespread and systemic, it is the height of folly.

This brings us to a key lesson of analytics: Sometimes the best decision supported by analytics is to maintain the status quo. Such a result shouldn't be considered a failure to produce. Rather, it should be considered one of the accepted outcomes of performing due diligence. An analyst might conclude that the current system is "close enough" to the best possible system and that any large-scale change simply isn't worth the risk. By definition, some predictions fail. The fallout of such failures needs to be considered part of the analytic decision-making process.

Small Steps Can Help

We've talked about a couple types of managerial innumeracy. There are certainly other types of innumeracy and logical fallacies that you should watch out for.[2]

Many have argued that businesses and organizations must understand the numbers and data that drive their business in order to survive. Certainly if your competitors are using analytics and you are not, you could be at a distinct disadvantage. Even in areas outside business, such as nonprofits, education, or healthcare, people have argued (and we agree) that driving more decisions with numbers and solid analysis will help these organizations better achieve their desired outcomes.

So, for the analytics movement to be successful, they must overcome managerial innumeracy. This may be a challenge for individuals and organizations as it may require them to pick up new mathematics skills and deal with data in a more logical way. The good news is that even a small step toward more numeracy can have a nice impact. You don't need to go to the extreme of being perfectly numerate to get benefits.

Analytics Is a Mindset

As a counter to managerial innumeracy and to fully embrace the analytics field, you must have an analytics mindset—a mindset where you are committed to using data in a logical way to help drive business decisions. This often requires the use of sophisticated models. To get the most out of these models, you need to understand the strengths and weaknesses of the models. You also need to be able to question the results from a model, be able to determine when a model has returned good answers, and understand how much confidence you have in each of these answers.

Gary Loveman, the CEO of Caesar's Entertainment Corp., summed up the analytics mindset best when he stated that "my job is to ask lots of penetrating, disturbing, and almost offensive questions." He is right on. To get the most out of your analytics projects, you need to ask tough questions and demand a lot from your models. In some cases, asking the questions and going through the process will teach you as much about your business as the final answer that comes out of the analytics model.[3]

You might wonder why we need to ask tough questions of our analytics models. Shouldn't we let the results guide our answers?

First, analytics models, like many other types of projects, can have bugs or design mistakes. By asking the tough questions and doing testing, you have a much better chance of ensuring that your models are correctly capturing the problem you want to study. Creating analytics models is in some ways very similar to software engineering projects. Anyone who has experience in software development will tell you that the first version of a new software product will have bugs—probably lots of them. That is nothing to be ashamed about. In fact, recognizing the likelihood of bugs actually leads to a faster road to clean and bug-free software. By releasing early versions with bugs, you can discover and actually fix problems faster. As a side note, even if you try to release the first version bug-free, you will almost certainly fail and, worse, your first version will be released much later, causing a major loss in momentum.

Second, analytics models have strengths and weaknesses. These strengths and weakness may not be initially clear even to the designer of the solution. Many assumptions are made when the models are built; and many of these assumptions are implicit, not explicit. By making one decision in the design, other assumptions unintentionally result. For example, in trying to predict how a new Starbucks in an area may impact the sales of another nearby Starbucks location, one might create a model that assumes that people would visit the store nearest their home. This might appear to be a reasonable assumption. However, this explicit assumption could be violated for many reasons. Suppose one store is near a park where people like to sit and drink their coffee; this might lead them to travel further to this store. An implicit assumption here might be that we are assuming that we can predict traffic based on where people live. Perhaps you have better predictive value by understanding where people work, where they take their kids to school, or where they go to the health club.

Third, you need to understand the models and the results. We have found that the more complicated the model, the less people are willing to ask questions. This is exactly the opposite of what you need to do. Just because a model is complicated doesn't mean it is correct. By definition, a complicated model is likely to have more hidden assumptions and hidden weaknesses. It may, in the end, give you insight that was impossible to find with simple models, but you cannot simply take it at face value.

People need to ask questions about whether the model is providing results that agree or disagree with what you expected. If the model agrees with your expectations, you should check to make sure that an analyst didn't somehow bias the result in order to please management. If the model disagrees with your expectations, don't discount your experience and intuition. The analytics may prove you wrong, but often your gut will help you identify problems in the analysis in general.

Besides the need for tough questions, it is also worth pointing out that developing an analytics mindset is not trivial in a larger organization. There are typically many forces at work in an organization that prevent the best analytical answers from being used. This book does not get into organizational behavior and how to overcome all these forces. However, we wanted to point them out so you are aware of them and know what to look out for.

Gary Loveland also pointed out that people often win out over ideas.[4] You can clearly see how this force could quickly wreck analytics projects: You deliver great analytics results that indicate that one direction is better than another only to see the idea overturned because someone with more power or influence wants to take a different direction. To make analytics projects work in an organization, you often have to work hard to use your fact-based results to win over key people.

Another well-documented problem is termed *confirmation bias*—the tendency to accept results that confirm your currently held beliefs and dismiss results that refute them. This can wreck analytics projects in many different ways. During the design of an analytics model, the designers may make assumptions that reinforce their currently held beliefs. When data is collected, especially manually, it may be collected in a way that confirms the output we want (even if we don't realize it). Alternatively, when results are presented, people may not question results that reinforce their preconceived notions or may dismiss out-of-hand models that contradict thoughts.

The term *confirmation bias* has typically been used in the area of politics. But, as Harvard Business School's Michael Schrage points out, there is often a lot of competition in politics, and ideas are quickly debated in the open. He also explains that this doesn't tend to happen within organizations. That is, once an organization gets a notion that one idea is better than another, confirmation bias can really kick in. In many organizations, it may actually be better for your career if you go along with existing ideas rather than fight the current. This will then lead to the development of analytics models that support or are manipulated to support the current view.[5]

To prevent confirmation bias and the many other biases, it is good to get in the habit of asking tough questions about the models and maybe also about yourself before accepting any model's results.[6]

The 80/20 Rule

The world and the organizations that operate in it are complicated. We need ways to separate the important from the trivial. The 80/20 rule helps us do this and is an important aspect of the analytics mindset.

The *80/20 rule* is a rule of thumb derived by the Italian economist Vilfredo Pareto. You will also often hear it referred to as the *Pareto principle*. The rule states that 80% of the outputs come from 20% of the inputs. Translated to business, this could mean that 80% of the sales come from 20% of the customers. Or 80% of the profits come from 20% of the products sold. Or 80% of the cost of the product comes from 20% of its raw materials. And so on. You can apply this to many aspects of your business. Keep in mind that it is a rule of thumb, not a mathematical law. It turns out, though, that it is quite often fairly accurate in a wide range of business applications.

If you didn't know the 80/20 rule, you might be tempted to think that improvements are linear. That is, you might think that if you have 10 projects, you get 10% of the total benefit from each one. Or, if you want to include 10 features in an analytics model, you might think that you will get 10% of the benefits from each of the features. The 80/20 rule warns that it is likely that 80% of your benefits in both cases will be from the top two items. As managers and analysts, we must determine which two.

One large company, Illinois Tool Works (ITW), has made the 80/20 rule one of its core principles for making decisions. The cover of ITW's 2012 Annual Report is actually devoted to the 80/20 rule. On the company website, ITW points out that the 80/20 rule says that 80% of their revenue comes from 20% of their customers. Therefore, the company uses this rule to stay focused on the most important customers. It also realizes that the rule is flexible and uses it in different contexts as well. The key thing the rule does for ITW is keep the company focused on the important parts of the business. ITW claims that it is the best in the world at implementing the 80/20 rule as a management practice. It believes this focus on the important 20% has been one of the reasons for its rapid growth since the 1980s. In 1985, ITW's revenue was $592.3 million (about $1.3 billion, adjusted for inflation). In 2012, its revenue was $17.9 billion. That is approximately 13%

annual growth (or 10% after inflation) over three decades. That is a strong endorsement of the 80/20 rule.

So even if you don't run your entire business by the 80/20 rule like ITW does, this rule is an important part of an analytics mindset. You need to be able to focus your efforts on the 20% that is going to drive 80% of the benefits. During any analytics project, you will be confronted with many different issues and decisions. You have to be able to decide which ones are important and driving the bulk of the results and which ones are merely trivial. For the trivial issues, even if you spend a lot of time working on them, the payoff will likely not be great.

Incorporating Variability into Your Analysis

An important consideration in many analytics exercises is the recognition of the fact that there might be uncertainty in many of the factors you are considering. Too often people simply look at a single number (typically the average) to describe some critical factor and miss the fact that this factor can actually take on many values, resulting in widely variable output as well. Sam Savage does a nice job of explaining what can go wrong in the process in his book, *The Flaw of Averages*. He warns people against the "Just give me a number" mindset. In many cases, just providing a single number can hide some very important information. For example, suppose you were asked to babysit my children for 5 hours. I could tell you that your expected payment would be $100. If I didn't tell you whether I would simply pay you $100 or enter you in a lottery where you had a 1 in 1,000,000 chance of winning $100,000,000, I would be hiding some useful information. In both cases, the expected payoff is $100, but in one case it's a sure thing, while in the other case you have some infinitesimal chance of striking it rich. You might decide to roll the dice, but certainly you should know what you are getting into before making the decision.

As you approach analytics projects, it behooves you to think about the key inputs to your models and recognize which have values that are relatively certain and which have a significant degree of uncertainty. Trying to understand how that uncertainty affects your conclusions is critical. This may involve using simulation techniques, what-if analysis, or other techniques.

The more serious error that can occur when you don't consider the uncertainty in your model is what Savage calls the "Strong Form of the Flaw of Averages" (the previous example represents the "Weak Form"). This error occurs when you don't realize that the expected result of the average is not equal to the average expected result. Suppose that you are opening a new Broadway show, and you don't know whether it will be a hit or a flop. With 50% probability, it will be a hit, and there will be demand for 3,000 tickets per night. On the other hand, there is a 50% chance that it will flop, and demand will be 1,000 tickets per night. Suppose you will sell tickets for $100 each and there are 2,000 seats in the theater. This would imply that you have a 50% chance of grossing $100K per night (flop) and a 50% chance of grossing $200K per night (hit...remember you can't sell more than the number of seats in the theater). If you wanted to arrive at one number, you would say your expected revenue is $150K per night. However, if the theater manager had simply asked for a single number for forecasted demand, you would have had to tell him 2,000 per night (50% chance of 1,000; 50% chance of 3,000). The manager might then (wrongly) conclude that the expected revenue per night was $200K.

A more colorful way of demonstrating this second error is to imagine an intoxicated man stumbling down the center of a busy highway. On average he walks down the center line and is therefore safe. In reality, he is stumbling into traffic and has likely been hit by a car. The result of the average (on average he is on the center line) does not necessarily equal the average of the results (the result is that he is hit by a car).

You Can't Just Use Accounting Data

At the start of an analytics project, we often ask whether the team has good, clean data. Their answer is almost always "yes." Once the project starts, however, it turns out that the data is not really clean. A lot of time is then spent cleaning up that data.

Were we intentionally deceived at the start of the project? We don't think so. Instead, people in companies often think that they have clean data because the data allows the business to continue to run and meet financial reporting obligations.

So, although the data may be clean for certain purposes, it does not mean that the data is ready to use in an analytics project. This is where you need to have a tough analytics mindset: You need to be able to recognize that data suitable for one activity (financial reporting) may not be suitable for an analytics project. This is not an easy leap to make. For example, people in the organization may not understand why data that works perfectly well in one area is not good for another area. Or people might be skeptical of creating yet another data set.

One of the problems you encounter with accounting data when using it for analytics is that it allocates costs in ways you might not want. In any system where you share a resource (a machine, a building, a truck, and so on) and have overhead costs (management, electricity, maintenance, and so on), the accounting system will systematically allocate the costs of these resources and overhead to individual products or departments. The method may be systematic for financial reporting, but it may not be suitable for making the decisions you want in your analysis.

For example, in a company that has three different departments and a management overhead cost of $1 million, a financial system may allocate that $1 million based on the total number of people in each of the three departments. There are two problems with this approach. First, the management may very well spend 75% of its total time and effort with the smallest of the departments. The accounting system shows that this department uses the smallest portion of the $1 million when in fact it actually uses the most. Second, by the time you get the data, you may not even see the $1 million broken out. It may be buried in the data for each of the departments, and you would have no way of knowing that there was $1 million in overhead allocated based on some previous assumptions.

This same problem often applies to products. In manufacturing environments, the cost of the plant and the overhead gets automatically allocated to products based on some measure—like units produced or direct labor hours used for the product. However, these costs may not even come close to the true cost to make the product. The true cost depends on how much direct labor they use, the resources they use, and the management time devoted to each product. So the accounting system may work for financial reporting needs, but it can lead you to make very bad decisions for the business. You may discontinue a product that shows a very high accounting cost but in reality is a low-cost product. When you discontinue this product, the overhead costs that were allocated to it don't go away; they just get moved to other products.

Managers have realized that this is a problem (but they often don't do anything about it). One solution is to create a parallel activity-based cost accounting system (sometimes referred to as ABC). Such a system works by understanding how much time, effort, and cost is devoted to each activity in the business. The activity could be a department or a product. This type of system often works well, but it can also be very time-consuming to implement because you must look at every specific activity within the organization.

Another solution is to make sure you look at unallocated cost data from accounting. This can usually provide you most of the information you need without requiring you to do a full ABC analysis.

In any case, the main point is to carefully question the data you are using so you can truly understand it. We will continue to elaborate on this in the next section.

Thinking Clearly About Data

Not All Numbers Are Data

Perhaps the first concept to make clear here is that not all numbers are data. This may seem a bit pedantic, but it is an important distinction to make because different types of numbers should be treated differently in analytics. Recently one of our authors was in a silly argument about a sports team with an old college friend. The friend complained that we were basing our argument only on *historical* data. To which we responded, "Is there another kind?" In truth, all data is historical. Data represents information that has been observed in some way. We cannot observe something that has not yet happened, so numbers about the future are not data; they are projections.

It is important to keep in mind that data is a means to an end; it is not an end in itself. Data can serve many useful purposes, but absent those purposes, data has no meaning. When we think about the "life" of data and its role in analytics, we can think of the following basic progression in the use of data:

1. Data allows us to understand what has happened and allows us to discern why.

2. Data enables us to discover the relationships between different variables.

3. These discovered relationships allow us to make projections on the future course of these variables.

4. Using these projections, we make educated decisions about our course of action going forward.

So what numbers are not data? As previously mentioned, we believe it is important to make a distinction between data and projections. Ideally, data is something that should be treated as facts (although we will discuss caveats to this shortly), and there should be little or no argument about the data itself. As a famous quote, typically attributed to U.S. Senator Daniel Patrick Moynihan, goes, "You're entitled to your own opinions. You're not entitled to your own facts." Projections are essentially opinions and should always be subject to debate. The future is unknown. When we present, discuss, or debate numbers about the future, we must always remember that this debate can be informed by data, but it is not about data. It is about projections. As we will discuss at length, the field of predictive analytics is dedicated to using data and mathematics to create reliable projections. Nevertheless, these are opinions about what will happen in the future. That is not to say that all opinions are equal. Some are well informed and reasonable and others less so. The task of discerning which is which is the responsibility of an analytics-capable organization.

Another example of numbers that are not data is statistics. This is probably a bit of a fuzzy area for people, but we believe it is important to distinguish the two for a few reasons. Rather than being data (information that has been observed), statistics are numbers that describe data. Statistics typically are based on formal mathematical definitions.

If this is the case, and if the data can be treated as fact, then in turn the statistics could also be treated as fact. In this sense, data and statistics are very similar in that they should truly be facts about which there is no debating. However, a few things can go wrong in this process. First, the data itself can fall short in a number of ways. (We expand on this in the next section.) Second, the statistics being presented may not be precise. For example, the most widely used statistic by far is certainly the average. Much of the time, the context provides you with enough information to know what this means, but in reality, the average is a statistic that can take on many meanings (arithmetic mean, median, mode, and so on). Finally, we have probably all heard the quote most often attributed 19th-century British Prime Minister Benjamin Disraeli: "There are three kinds of lies: lies, damned lies, and statistics." The reason this is such a memorable quote is that all of us have had the experience of being presented statistics that we believed, or knew, to be inaccurate, incomplete, or misleading. When creating statistics, there is a lot of discretion on the part of the analyst on what statistics to use, which data to include, and how it should be presented. Because of this, the result can imply directly or provide reason to infer an inaccurate conclusion. Sometimes this can be due to the analyst presenting the data in a way that supports or confirms a preconceived bias. However, often this happens unintentionally, either because the analyst does not recognize the bias or because of simple human error.

The short version of all this is that you should keep in mind that data and statistics are different and should be treated differently. Statistics, like projections to some extent, should be open to debate in particular when you are being asked to make conclusions or recommendations based on those statistics, which is most likely the case. (Otherwise, why are you being presented these statistics?) At the very least, you should find answers to the following questions:

- How did you decide to present these statistics? Is there any ambiguity in the definition and any bias being shown in the choice?
- What data is being included, and what data is being excluded? Why?

- Are there any other statistics that might lead you to other conclusions?
- How accurate is the data? (More on this next.)

Can We Trust the Data?

As mentioned above, ideally data can be treated as facts. However, outside of some very limited, controlled experiment types of situations, we expect that it will not take much to convince most readers that this is not typically the case. In most real-world cases, data accuracy is a persistent and pervasive problem. This is true for potentially many reasons, and we will highlight only a few here. The reason we do so is that when you are going to use data for analytics purposes, you should consider the possible sources of errors in that data so that you can have the proper level of confidence in the results.

Data Capture

One step in the analytics process that can result in errors is the data capture process. In almost any modern organization, the majority of data will reside in digital form. Gone are the days when most data was filed away in manila folders and steel cabinets. Because of this, it is likely that you might have too much faith in these numbers. More and more data is being automatically captured by sensors or other automated systems, and we often presume that there are not bugs in the software code and there are not machine failures, so this data can be expected to have a high degree of accuracy. However, it is worth thinking about how much of the data we review in digital format had a quite manual start and is prone to human error.

As a simple example, one of the monthly bills received by one of our authors is addressed to "Derrick Nelson," not to Derek Nelson. The bill itself is clearly generated from a computerized system, so how did the misspelling come about in the system? Most likely, at some point he filled in a paper form with his information, and someone had the job of entering that data into the company's system. A number of errors could have occurred at this point. Derek could have entered the data incorrectly (unlikely in this specific example as he has been spelling his name for quite a few years), the clerk could have misinterpreted his poor penmanship (possible but seems unlikely, given the differences in multiple letters and word lengths), or the clerk could simply have entered the data incorrectly for some other reason. In this case, we suspect that the clerk read our information and then looked up at the computer screen, remembered the name, and wrote the expected spelling.

The incorrect name example is rather benign, but you can certainly imagine situations where this type of error could have more significant impacts. In our professional experience, we have done many projects in the trucking industry. These projects have often been focused on the question of whether by using some different algorithms it is possible to create better routes and schedules in order to reduce miles driven and ultimately the cost of operations. Typically the first step in this type of analysis is to use a set of historical shipments and see if, in retrospect, we could have done better than the company actually did. Critical to looking at this problem is to have data on both the weight (pounds or kilograms) and the volume (cubic feet or cubic meters) of the shipments. This is important because the trucks will have limits on both weight and volume that must be respected as routes are designed. In nearly every project of this type that we've seen, there have been shipment records that were missing either a weight or a volume (or both!). How does that happen? Most likely, someone in the daily operations has responsibility for recording and entering this information into the system. It may be that that person was so busy in fulfilling her primary responsibility of making sure the shipments were made that she did not have time to record the data. Another possibility is that she knew a certain piece of data was unimportant for her current purpose; for example, she might have known that a certain truck was going to reach its weight limit not its volume limit, so in that case recording the volume may have seemed unnecessary. There may also be many other reasons, but regardless of which ones are correct, the data was not captured, and therefore you would need to make an assumption about the missing data when doing your analysis.

Believe it or not, the trucking situation just described is actually better than other situations we have seen. Usually there is simply no data entered for the volume; in this case, we would easily recognize this and know that we have to make an assumption. In a worse situation, however, the system does not actually allow an operator leave this data field blank, so finding herself needing to put in some value, she simply enters a 1 so that the system accepts the record and she can move on. This is potentially much more harmful because someone using this data, without inspecting it for reasonableness, would assume this it was a "real" data point. This could possibly result in designing routes that are not physically possible because the data would lead you to believe the shipment was smaller than it actually was.

Needless to say, understanding where and how certain data is captured should be something to consider when evaluating how much trust and scrutiny you may place on it.

Almost the Right Data

Another area of concern when evaluating your level of trust in data is whether the data truly represents what you are looking for. It may be close to what you are looking for, and often exactly what you are looking for does not even exist. Probably the most common case of this within companies is in the area of demand data. Virtually all companies face the problem of projecting future demand for their goods and services. To do so, they almost always look at historical data to inform their projections. They look at past sales and call that their historical demand. The problem is that in many (or most) cases, sales and demand are not the same. They are certainly related, but they are not the same. The difference is that we are not capturing demand, and we have limited, if any, ability to capture demand that existed that did not translate into an actual sale. Demand may not result in sales for many reasons. Perhaps a consumer goods company ran out of stock of an item. Perhaps a services company did not have a sales force large enough to reach all the demand that existed for the company's services. Perhaps there is not enough awareness of your goods or services, and people who otherwise would have purchased your product could not because they didn't know it existed. No matter the reason, you risk making improper conclusions about possible future demand when you mislabel the data you have as historical demand when really it is a related but different piece of data:

historical sales. Many companies are well aware of the concept of lost sales, or "shadow demand." However, in our experience, it is still extremely common for companies to treat historical sales and demand as synonymous in their analyses.

We have highlighted the most common example of this type of data concern, but there are certainly other cases where the data available is close but not exactly what a company is trying to measure. Another possible example in the world of sports might be the data on home runs for a baseball player. Certainly the home runs per at-bat is well known for all baseball players and would be a very useful metric to use in projecting future home runs that player may hit. However, what this misses is that not all baseball stadiums are identical, and different weather patterns may affect whether an at-bat results in a home run or not. In this case, perhaps the best data would be the velocity, direction, and trajectory of the ball after it leaves the player's bat; if you collected that data, then knowing the stadium configuration, wind, and physics would allow you to project the likelihood of a home run under various conditions. Given the extensive analytics performed around Major League Baseball, for all we know this data is already being collected (or will be in the near future).

Data That Can Go into Multiple Buckets

Another area to watch out for when putting your trust in data is whether there are cases where there is discretion about what bucket some piece of data falls into. Let's illustrate this with a simple example. Suppose you have a company that sells furniture. In your business, it is common to discount in order to win the customer's business. The salesperson may have discretion in discounting the sales price of the furniture, or he may have discretion in discounting (or providing for free) the delivery of the furniture. In looking at the data across different salespeople, stores, or regions, you would need to be careful about looking at data on sales price or delivery charges independently. Most likely, you are mainly concerned about the total revenue from each customer. Focusing on sales price or delivery charges independently can make sense, but you must recognize that there is a level of discretion and arbitrariness in how the salesperson decides to allocate the revenue.

Simple Tests You Should Do with the Data

It can often be intimidating to work with a new data set. And for large data sets, it can be difficult or even impossible to simply look at the data and see if there are problems. Unfortunately, because of this, many people tend to just assume that large data is correct and do nothing to check the data in general.

If you are going to feed data into a system and use it to make decisions, you can do some simple checks that may find most of the problems. These checks are almost too simple to state, but we have seen too many people skip this step in their analysis and only later find problems with the results. These simple checks should be built into any process that involves analyzing data. You can do these checks manually or automate them. You should just make sure they are done.

This list is not exhaustive but should give you an idea of what to do. For each column of data:

1. Find the *average* (start with the *mean* and then maybe look at the *median* and *mode*) and see if it makes sense.
2. Find the *standard deviation* in the data and see if the amount of variability is what you expect.
3. Find the *minimum* and *maximum* values and see if they make sense.
4. Related to step 2, if the data is supposed to be positive, *check for negative numbers*.
5. Check for *missing* values.
6. Check for *non-numeric* values in a column that is supposed to contain numeric data.

You should also do some simple checks across columns of data:

1. If a column is derived from two others, make sure it works. For example, if the revenue column is supposed to be price multiplied by units, make sure it works out that way.
2. If two columns are supposed to have exactly the same data, make sure they do. (It seems strange that you would get data where two columns have the same data, but it happens sometimes.)

3. Determine appropriate *ratios* between the columns, calculate those ratios, and then run the simple checks on those ratios. For example, if you have a column for total units and another one for weight, take the ratio to determine the average weight per unit for each record. Then see if that weight per unit makes sense. This is somewhat subtle but helps find hidden problems.

Finally, you should build a few graphs with the data to help understand it. Common graphing includes:

- Plot the data as a *histogram* to see it that makes sense.
- If your data has dates, plot a *time series graph* to show the various columns over time—sales over time, units returned over time, and so on.

Making Do with What You Have

A lot of this chapter may make you think it is impossible to have clean and usable data. In one sense, knowing this may make your lives easier in the short-term: If your data is in such bad shape, you can avoid doing an analytics project. But this kind of thinking is rarely good for your career or your organization. Instead, the previous sections of this chapter have hopefully given you ideas on how to make the data better.

But, more importantly, a good data analyst must also be able to determine when the data is good enough for the analysis. This takes us back to the 80/20 rule: You are going to get 80% of the value in a data study from 20% of the inputs. So you might focus your data cleaning efforts only on the 20% of the data that is going to drive your results. Or, said less strictly, the 80/20 rule strongly suggests that your data inputs don't have to be perfect. This is why historical sales may be a good approximation for demand.

You also have to balance the trade-off between making decisions with data that is less than perfect and making decisions without the data. In other words, this chapter doesn't give you an excuse to do nothing because your data isn't perfect. It never will be. Instead, this chapter is meant to balance your enthusiasm for new analytics tools with the need to make sure you have the right mindset to use those tools.

Problems with Bad Data from the Start

Stephen Budiansky wrote a book called *Blackett's War*, which highlights the role that English scientists played in defeating the German U-boats in World War II. Budiansky also cites this as the birth of the field of operations research—which is applying the ideas from math and science to management. Operations research is of interest to readers of this book because a lot of people consider this the forerunner to the analytics movement. In fact, the scientists who worked on the U-boat problem had a varied background (similar to how the field of analytics has pulled together many different disciplines) and attempted to use data to solve management problems. So, although at times it seems like analytics is new, it is nice to get some perspective and realize that much of the groundwork was laid for this movement many years ago.

Budiansky's book gives two great examples of how bad data, bad assumptions, and bias can result in bad decisions.

The first example from the book centers on the then newly installed radar and anti-aircraft guns stationed in England to protect the city of London. Anti-aircraft guns had been around for a long time, although they had never been very effective. It then dawned on scientists to link the newly created readings from the radar to the anti-aircraft guns to improve their aim. After some initial glitches, the anti-aircraft guns did aim better and were able to shoot down more bombers than ever before. To protect a city like London, however, gun stations were placed at the coast and some closer to the city as well. Early reports from the people running the guns indicated that the guns stationed at the coast were getting twice as many kills as those next to the city. Since the guns were the same and radar was still fairly new at this time, the scientists hypothesized that perhaps radar worked better over the sea than over land. If this were true, they should move the guns from London out to the coast. However, before they moved guns, they wanted to understand the data better.

After some further investigation, the leader of the scientists discovered the true source of the problem, though. The kills recorded over the land were easily verified by finding the wreckage of the downed bomber plane somewhere around the city. This negated any overinflated estimates of the crews working the guns near London. The kills at the coast could not be easily verified. Most of their downed planes sank directly into the sea. Therefore, their kill numbers could not be verified and were simply taken as reported. Since the objective of the crew was to get as many kills as possible, their bias was to confirm as many as possible. Once the scientists looked deeper into the data, they presumably saw that the gunners near London reported as many kills as their colleagues at the coast; however, their numbers were cut back after all physical confirmations of kills in their areas were completed. By recognizing the confirmation bias, the scientists prevented the moving of guns away from London to the sea—which according to *Blackett's War* would have actually had dire consequences.

The second example has to do with measuring the effectiveness of long-range Allied bombers hitting their targets within Germany. This measurement was important in order to help determine the success of the bombing campaign, which was a critical part of the war strategy. The bombers reported the success of hitting the targets personally. (This tells us we need to watch out for confirmation bias from the start.) But without data, the military leaders had no choice but to rely on the reports from their people in the field (the bombers, in this case). At one point, they were able to obtain aerial photos after bomb runs. What they found from these goes to show how bad confirmation bias can be: They chose to analyze just those bombers who had reported hitting their targets. Their results indicated that 80% of those did not even come within 5 miles of the target. Presumably, the other 20% were within 5 miles, but it is not likely that many of those hit the target either. This is shockingly bad, but it is not that surprising. But, don't just think about this analysis an example of bad data. It also impacted strategy. There was a debate within the military about how to allocate the long-range bombers. Should they focus their effort on hitting targets in Germany (and destroying production capability and eroding their ability to fight)? Or should some planes be diverted to anti-submarine activity (getting rid of the U-boat menace, which was sinking many cargo ships coming from the United States to the

United Kingdom and thus delaying the possible launch date for D-day because the allied forces could not get enough men and material staged in England)?

Blackett's War does a nice job of highlighting some of the organizational challenges of implementing analytics solutions. The military leaders were not easily swayed by data, and the people pushing the data analysis were civilian scientists (who were working with the military as part of the team but not part of the official leadership). This also shows that you need to be mindful of organizational factors when pushing analytics solutions. Both of these examples should give you pause when you are using data that is subject to confirmation bias.

The Rise of the Data Scientist

Because of the importance of the analytics mindset to the analytics movement, we shouldn't be surprised that a new job title has quickly arisen: *data scientist*. This job title has become very popular. Companies are looking to hire data scientists, and many people are describing themselves as data scientists on LinkedIn and other professional networking sites.

With the confusion around the definition of *analytics*, it is not surprising that there is also confusion about what a *data scientist* is. We see the same problems as with the term *analytics*: Companies and people make claims about what a data scientist is but are only covering one or maybe a few aspects of the job. For example, someone may only know descriptive analytics but claim to be a data scientist.

Instead, we should embrace a more holistic definition and define a *data scientist* as someone who has an analytics mindset and is committed to the full field of analytics—someone who is willing to use data to help make better decisions. We will not limit it to a set of skills or a specific part of analytics. Those in this role must be aware of all the different types of analytics. A data scientist doesn't need to be an expert in all the different areas of analytics but should know about all the areas and be willing to use them if a problem calls for it; this is where the "scientist" part comes in.

So, a data scientist isn't just a new trendy name for old professions. Instead, it is a recognition that the analytics field is broad and that you need to have a particular mindset in order to embrace it. For example, if you are an expert in statistics but want to be labeled as a data scientist, you should embrace the full field of analytics and its relationship to you. You can be a deep expert in one part of analytics and still be a data scientist. On the other hand, if a statistician refuses to embrace or learn about other areas of analytics, it would be difficult to label that person as a data scientist.

The new title *data scientist* also shows that the market is demanding people who are adapt at many aspects in the field of analytics.

The next section of the book discusses the tools that make up the field of analytics.

Part II
Analytics Toolset

4

Machine Learning

Introduction to Machine Learning

Before we dive into detailed accounts of descriptive, predictive, and prescriptive analytics, we need to pause and introduce some of the key algorithms of machine learning. (Don't worry, we wrote this chapter as if you have never heard of machine learning before.) Most of the machine learning algorithms discussed in this chapter are used for predictive analytics. So we are stealing some of our own thunder by covering machine learning now. However, the reason we introduce machine learning now is because this is a relatively new field of study and has applications to all three types of analytics.

A lot of exciting applied research is happening, with frequent advances, in machine learning. Much of what is new in the field of analytics is coming from the research of machine learning. For example, email spam-filtering systems, facial recognition in Facebook pictures, and Amazon's personal recommendations are all examples of machine learning algorithms at work. Since this is a relatively new topic, it is not yet frequently taught in business curriculums. So, as a manager, this is likely to be the area in the field of analytics that you have had the least amount of exposure to. This is also what you may hear the most about when speaking with others about analytics. This chapter is meant to give you the information you need in this subject area.

If you've never heard the term before, *machine learning* may sound like the stuff of science fiction—computers teaching themselves. Or perhaps you've seen machine learning mentioned in an article about Google's driverless cars. It likely sounded very advanced and like it was meant only for the most dedicated scientists.

The reality is that many of the machine learning algorithms are very accessible. Every manager needs to know the basics of machine learning. This material will eventually be widely taught at the university level, and there will be many readily available tools.

In some sense, machine learning grew out of what used to be called *data mining*. Of course, with the rise of databases decades ago, *data mining* (which refers to looking for information in the data) wasn't a bad term. As the data has grown, however, so have the algorithms that are used to extract the data.

We will eventually place the different machine learning algorithms into either descriptive, predictive, or prescriptive analytics categories. But for now we need to go over all the different terms and types of machine learning algorithms, regardless of the analytics category. You will hear people use different terms to describe machine learning. And, since it is relatively new, important, and a big part of predictive analytics, you will also hear a lot about it. From a manager's point of view, you need to understand these terms and where they fit into the larger analytics movement.

One caveat for now: No explanation of machine learning that we provide will be universally accepted. Experts in this area will define aspects of machine learning in different and contradictory ways. Our goal is to give you enough context so that if someone starts a conversation with you on this topic, you are not starting from nothing.

Let's start with data mining. Over time, the term *data mining* has morphed into *knowledge discovery*. At the same time that *data mining* became a popular term, researchers created the field of *artificial intelligence*. The term *machine learning* then grew out of artificial intelligence.

Machine learning grew from the fact that researchers realized there was no way to tell a machine everything it needed to know. Researchers saw that they needed to create algorithms that would allow a machine to figure out information on its own, given a set of data to start. Machine learning is a multidisciplinary field that takes techniques from statistics, probability, computer science, physics, and more. Although some researchers with a statistics background use the term *statistics learning* (instead of *machine*), a lot of the research and writing on machine learning seems to be coming from people with a computer science background who prefer the term *machine*.

A couple trends have shaped the field of machine learning (from a manager's perspective):

- Some of the same techniques that were being used in data mining and knowledge discovery were being used by machine learning specialists. That is, a technique that would help find patterns in a database could also be used to teach a machine a pattern. So, in a sense, these fields have merged.

- With the abundance and size of data, the emphasis in the field has shifted from applying ever-more-advanced techniques to small data sets to applying simple techniques to large data sets. This means that most of the algorithms being used are accessible to normal business users.

The language that machine learning experts use can sound overly technical, but the concepts can actually be quite simple, and with a little explanation, they can provide a good basis for understanding.

First, machine learning algorithms are split into two categories: supervised learning and unsupervised learning. With supervised learning, the data set you are working with has information on the desired output or answer—that is, it has what you want to predict. For example, if you are trying to predict a consumer's purchase patterns, based on website page views, the supervised data will likely have a column indicating whether the person actually purchased an item. Or, if you are working with medical data to determine the efficacy of different treatments, you might have a column of data that indicates whether the medical condition improved. Or, if you are attempting to recognize handwritten text, your data will have various handwritten numbers (which could be quite sloppy), with the actual number that is trying to be expressed coded into the data.

So, in all three cases mentioned above—consumer's buying patterns, efficacy of medical treatments, and handwriting recognitions—the machine learning algorithms look at all the data and try to predict the which factors led to a purchase, to a successful medical result, or to a successful reading of handwritten text. By having the answer within the data, you are "supervising" the learning by guiding the algorithm.

The data set you use when fine-tuning an algorithm is called the *training data*. That is, you use this data to train the algorithm how to make predictions. Despite the initial inclination to include all available data as training data, researchers quickly realized that if you give an algorithm all the answers, you don't have a reliable way to test the algorithm (that is, you will know how well the algorithm fits the data but not how well it will predict when it doesn't know the answer). If you use the same training data to test your algorithm, you may find that the test results look good because the algorithm may capture every nuance and outlier of the data set. The algorithm over-fits the model and does a fantastic job of predicting the results from the training data. But when you try to use it to predict new observations, it performs poorly. So, when building and fine-tuning your algorithms, you want to hold back some of the data and use it later as your test data. That is, after the algorithm is set up to make predictions with the training data, you feed it the test data (without allowing it to use the answers in the test data set) and see how it does. The algorithm predicts a value, and you compare that to the answer you know. This method provides a better validation of your model for future deployment.

This points to a good lesson for analytics in general: Be careful to build good and unbiased test data so you can test your analytics models. It is much better and cheaper to test your models before you actually try to implement them.

With this background, it now makes sense to talk more about the "learning" part of machine learning. The simple idea to grasp is that the machine learns from the training data you give it. But the more advanced idea is that the learning process can be ongoing. That is, the algorithm can learn from the initial data, and then as new data arrives, it can make further predications or observations about that data as well. If you take the time to set up an algorithm to provide feedback on how well it is performing, then new observations are automatically made part of the training data going forward. The algorithm automatically updates itself with new data. You can see the appeal of this. In a data-rich world, we don't want to have to rewrite algorithms to take advantage of each new data discovery. We would like our algorithms to "learn" and virtually adapt themselves. But even without this feedback loop, machine learning algorithms can be powerful.

For example, earlier in the book, we talked about an online consumer finance business that created various algorithms that would compete as new customers arrived at its site. Embedded in the algorithms was logic to determine which webpages would be shown to which customers and what promotional offers would be made. The management team didn't want to decide what to show or offer a specific customer; instead, they wanted the algorithms to decide (and presumably make much better decisions). In practice, when a customer arrived, a certain amount of information about the customer would be known (demographics, times to the site, referring site, other sites visited, previous business, and so on). Each algorithm is given a certain number of credits—think about this like Monopoly money—and the algorithms would then start a process of bidding on the customer with those credits. The algorithm with the highest bid would get to show the customer its choice of webpage and its offer. The ongoing learning part was that the algorithm received more credits to use for future bidding if the customer purchased a product (a reward for good performance). This would give the algorithm more information about the value of different customers. The mechanism also prevented poorly performing algorithms or offers from continuing: They would eventually run out of credits.

Let's look at each of the major algorithms that make up machine learning. There are many other algorithms and many variations of the algorithms we discuss. This list will give you, as a manager, a starting point for your research and give you a reference point for other algorithms in the future.

Supervised Machine Learning Algorithms

There are many types of supervised machine learning algorithms—so many, in fact, that they can be broken down into subcategories. One of the main subcategories is classification algorithms. K-nearest neighbor and decision trees are two of the most popular classification algorithms. Supervised learning algorithms also include recommendation systems and even regression analysis.

We like the fact that regression analysis is part of machine learning. It is a key idea from statistics and is widely taught, widely used, and even available in Excel. It shows that machine learning is not some magical set of equations available to only an elite few.

Classification and K-Nearest Neighbor

We like to highlight k-nearest neighbor because it shows both the power of machine learning algorithms (why they get so much hype) and how accessible they can be.

We use complicated terms like *machine learning, classification algorithms*, and *k-nearest neighbor*, so you might be surprised how simple the concepts here are. Let's look at the simplicity of the k-nearest neighbor algorithm with an example. Say you have a data set of 5,000 people who own cars. The data set shows, for each person, the income of the person, the age of his or her car at the start of the year, and whether he or she bought a new car by the end of the year.

This is a supervised algorithm because you want to be able to predict whether a person (the new data point) will buy a car this year, based on his or her income and the age of the person's current car. The training data contains exactly this information.

To predict the likelihood that a new potential buyer will actually buy a car this year, the k-nearest neighbor algorithm takes a new data point (a potential car buyer) and calculates the distance from this new data point to every point in the training data (in this case, the information around the 5,000 people mentioned previously). The distance in this case is a combination of how near each attribute of this new data point is to the attributes of the training data set (the 5,000 data points you have). For example, if your new potential buyer has an income of $45,000 and her car is 5 years old and one of the previous 5,000 data points has an income of $65,000 and his car was 6 years old, you simply calculate the distance between the point (45000, 5) and (65000, 6).

Once you make this calculation between the new point and every other point in the training data set, you simply find the k closest points (a software package does this for us). Now you see where the k in the algorithm name comes in. And you can see where the *nearest neighbor* part comes in, too: These are the points nearest to the new point. k could be 1, 5, 7, or any other number you choose. Let's say you set k to 7 for this analysis. Once you have the 7 nearest points, you will also know whether each of these people bought a car. Then, to find an answer, you simply take a vote of those 7 data points to see whether they bought a car. If 4 or more of these 7 people bought a car, you will predict that the new person will buy a car this year. You are just taking a majority vote.

Figure 4.1 shows a sample of 10 records for this type of analysis. This example shows what happens when you test your results on another set of data with 5,000 records—which would represent using your results to test the next 5,000 customers that arrive. In this case, the model predicted the right outcome about 75% of the time.[1]

Person	Income	Age of Car	Bought New
1	53400	11	Y
2	55100	5	N
3	19300	3	N
4	36000	8	N
5	218600	4	N
6	66900	3	N
7	66500	3	N
8	38300	7	Y
9	99100	7	Y
10	68300	6	N

Figure 4.1 Sample of the 5,000 Records for Potential Car Buyers

In another case, the training set of data is a large set of handwritten numbers or letters, with their corresponding actual values. (Think of this as the answer key.) To translate the handwritten numbers into something a machine can understand, you center and scale the handwritten number and overlay a grid on top of it. For example, Figure 4.2 shows a handwritten 3 with a grid that is 10 rows by 6 columns. This grid has 60 different squares. For each square, you (or, more likely, a computer program) assign the square a value between 1 and 0 to indicate the percentage of the square that is covered with ink. In the example in Figure 4.2, row 2-column 2 would get a value of around 65%, row 3-column 3 would get a 0%, row 3-column 5 (highlighted in the copy of the picture on the right) would get 96%, and row 5-column 4 (also highlighted on the right) would get 100%.

You would continue to assign these percentages for each of the 60 squares. Now, you can think of each of the 60 squares as attributes of this number. Once you've done this for the training data, you take a new handwritten text value and determine the value of each of the 60 squares for it as well. Then, the k-nearest neighbor algorithm finds the k closest neighbors to this new text value and takes a vote to see what the neighbors think the new number might be. For example, if $k=11$ and 7 of the closest neighbors are the number 3 and the 4 others are 5, then the result would be to predict a 3.

For this problem, in practice, the grid can be much larger. Note that the work involved in scaling and aligning the numbers is also not a simple task.

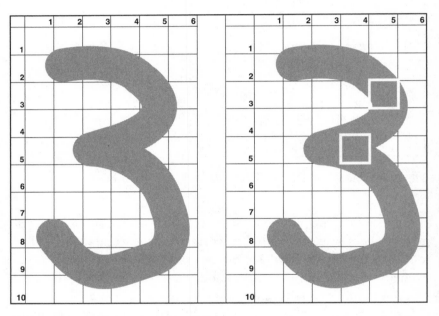

Figure 4.2 Grid for Text Recognition

There are a few more points to make with this example. First, you can see that it may require a lot of data; every single letter or number in the training database could have 10,000 points of information in a full-scale application. Second, you need to keep this full database around so you can check new letters or numbers as they arrive. This means that this algorithm could consume a lot of memory and must perform a lot of calculations each time it is used. But you can also see where the learning part of this algorithm might come in.

If the program has a way to get feedback on its results, it can continually grow the training data set with new information—making the algorithm perform better as time goes on. Google is actually great at this: It gathers more and more information to continually improve its searches.

Now that you've seen how this algorithm can recognize text, it is probably obvious that the same technique might be applied to facial recognition as well. For example, if enough people upload a picture of you on Facebook and tag you, then Facebook can turn those pictures into a big grid with various colors and maybe even some data off to the side that measures the distance between your ears and nose or other features. Now, if a new, unidentified photo is uploaded, Facebook can turn it into a grid of data and run the k-nearest neighbor algorithm against its data to make a guess about whose faces appear in that picture. In addition, it's possible that the engineers at Facebook have narrowed the search further by recognizing who was the source of the uploaded picture (that is, it is very likely that many of the faces are friends in your network).

K-nearest neighbor algorithms have also been used to detect email spam, identify credit card or other types of fraud, aid in medical diagnosis, predict responses to micro targeting, perform Twitter sentiment analysis, and much more.

K-nearest neighbor is a simple yet powerful machine learning algorithm.

Classification and Decision Trees

In management circles, decision trees have been around for some time as tools for making better decisions. Machine learning, in effect, uses the term in almost the opposite way: In this case, a decision tree helps you divide data to make predictions. To clear up or prevent confusion, we'll cover a traditional decision tree first and then go into the differences between that and a machine-learning decision tree.

A traditional decision tree is used to help make complex decisions where different outcomes or moves by competitors can impact what you should do. For example, the simple decision tree shown in Figure 4.3 can help you decide if you should invest in a new product, invest in an existing product, or do nothing. The results depend on whether the economy turns out to be good or bad next year and then whether your main competitor releases a new product or sticks with its old one. As you can see, this traditional decision tree is constructed from left to right. Each square block (just one in this case) represents a decision with different branches for the potential choices, and each circle represents a potential branch where several outcomes could result. If you follow the top branch all the way to the right, it shows that if you release a new product, the economy turns out good, and the competitor releases a new product, your net profit will be $80 million. If the competitor sticks with its old product, however, your profit will be $100 million. On each of the branches following the small circles, you see the probability assigned to that outcome. You can see in this tree that there is a 40% chance for a good economy and a 50% chance that the competitor releases a new product.

After you construct the tree and assign a value to each of the branches and probabilities of different potential outcomes, you can construct the expected value (the value of the outcome multiplied by the probability that the outcome will occur) of each of your decisions, which you see on the branch just to the right of the decision point. In this case, the expected value of $53.5 million is the best choice corresponding to investing in the existing product.

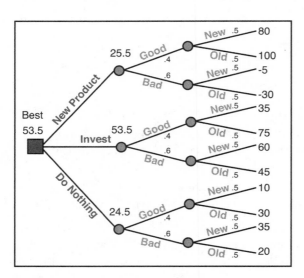

Figure 4.3 Traditional Decision Tree Example

Traditional decision trees can obviously get much more complex than this. Nonetheless, you can see from this example how you can now try different values at the end of the tree and apply different probabilities for outcomes. In this case, if the chance of a good economy were 80% (instead of 40%), the decision would change to releasing a new product.

You can see the value of such a decision tree: It provides a structure for making a decision, and it forces you to think about the probabilities of each outcome. Of course, the data could be manipulated (maybe unintentionally) to get a desired outcome. But if you use this as a tool to explore different probabilities and discuss the risks of different situations, you can arrive at better decisions than if you didn't take this approach. And, if you make a lot of similar types of decisions, you may also become more systematic in the probabilities you assign to different branches.

We've been looking at the type of decision tree you may have encountered in business school or a decision sciences text. The machine-learning version of decision trees is very different. In some ways, it deploys the reverse process of a traditional decision tree. The traditional decision tree has you design the tree first and then fill in the appropriate values. A machine-learning decision tree, on the other hand, starts with the data and uses the learnings from the data to decide how the tree should be designed.

Let's illustrate this with the same example used in the section on k-nearest neighbor. Once again, you have 5,000 potential car buyers, their income, age of car, and whether they bought a new car. You again want to be able to predict what a new potential buyer might do, but this time using a machine learning decision tree rather than k-nearest neighbor.

So, in a machine learning decision tree, the algorithm systematically starts with a particular column of data, splits that column into two or more branches, and then from each of those points, splits other columns into two or more branches. It continues to do this until it has exhausted the data set. To read the prediction, follow the decision tree to the end. The algorithms are sophisticated enough so that they branch on the most important column first—the column that has the most predictive power. You can also set the parameters so that the tree has fewer or more branches.

After the tree is designed, when a new point enters the system, it simply follows the tree to the end to determine which prediction applies. You can see the output of the machine learning decision tree in Figure 4.4. To read it, start at the top. The algorithm has determined that it is best to split the data first by the age of the car and then whether the age is 4 or less or greater than 4. If you look to the left, you see that it then splits on incomes of less than $124,300 and predicts a "no." You can follow the other branches down as well to see how their predictions result. The numbers in the final box show how well the algorithm did with the test data in terms of predicting the right outcome. (The first number in the final box shows how many outcomes it correctly picked, and the second number, after the slash, shows the incorrect predictions.)

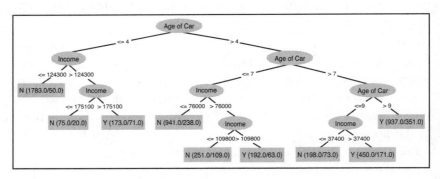

Figure 4.4 Decision Tree for Machine Learning[2]

In this case, the result of the machine learning decision tree was about as accurate as the k-nearest neighbor algorithm.

But, there are a few things to point out about this algorithm:

- The result of this algorithm can be expressed as a rather simple set of rules. If you have data for a new person, you simply need to take that person's data and follow the decision tree to the end to see what you would predict. You could write a very simple program to do this. Contrast this with the k-nearest neighbor approach, where you always need the full training set to predict a new outcome.

- The machine learning algorithm does not require numeric data. It can create branches based on text fields. If you added a column of data for a male or female buyer, the algorithm would decide when it was appropriate to branch on the gender of the buyer as well. If you have more than two text values, it would branch on as many as you had.

- The machine learning algorithm also gives you insight into your data. Figure 4.4 shows that when the algorithm ran, it chose the age of the car as the most important variable. You can also see how the algorithm branched on various ages of cars and income levels. This immediately provides additional insight into the data.

Recommendation Systems

You are likely familiar with recommendation systems from companies like Amazon, Netflix, or Match.com. These systems work very similarly to the k-nearest neighbor algorithm. They look at the things you like or have rated, find people who have rated items similarly to the way you have, and then recommend to you something that those who are similar to you like but that you have not yet rated or bought.

For example, if I have rated 10 movies the same way you have, and I really liked an 11th movie that you haven't rated, then the system would recommend this 11th movie to you. The idea is that if we had similar tastes on the first 10 movies, we are likely to agree on the 11th as well.

The recommendation system algorithms work similarly to k-nearest neighbor by finding the distance between my ratings and yours. So, if I rate the first movie as a 2, the second as a 5, and the third as a 3, the algorithms look for other people's ratings that are close in terms of distance.

These recommendation systems do have some complications that pop up, however. You should watch out for these issues in other types of projects as well. One issue is that different people use a different internal scale when they rate things. (You may recall an episode of *The Simpsons* in which Homer had a brief job as a food critic and said, "this restaurant gets my lowest rating EVER: seven thumbs up.")

For example, on a scale of 1 to 5, I may rate things that I really like as a 4, things I really dislike as a 2, and everything else as a 3. You may rate things evenly between 1 and 5. Others may rate everything as either a 4 or 5. So, even with what seems to be an objective scale, people will use the scale in different ways. This issue can be addressed by something known as the Pearson coefficient.[3] The math is a bit complex and beyond the scope of this book. However, all you really need to know right now is that the Pearson coefficient is used to normalize people's ratings and tries to find people who rate items in the same order. For example, if I rate three movies as a 1, 3, and 5, and you rate the same three movies in the same order as 2, 3, and 4, the Pearson coefficient scales these values in a way which indicates that our preferences may be similar. You may also run into this issue when you analyze survey data where you ask the respondent to rate service or some other metric.

Another issue is that the overall number of items that any one person has rated may be relatively small compared to the total number of items available to rate. You don't want all the unrated items to have a large influence. And, you don't want the fact that two people just happened to rate only one thing in common and rated that the same to have a large influence. Although the specifics are again beyond our scope here, it is useful to know that the issues around sparse data can also be addressed with different techniques, such as the cosine similarity.[4]

It is again easy to see here where the "learning" part of machine learning can come into play. First, as more people rate different items, the training data grows and gets better. You may also be able to track over time whether two people's preferences continue to align or whether they begin to diverge. Finally, if you can get feedback from people on the recommendations themselves, you can further adjust the process.

Regression Analysis

The last supervised machine learning algorithm we cover is regression analysis. It is a bit strange that this gets lumped into the category of machine learning since it has been around much longer and is well established as its own discipline within the field of statistics. But it definitely shares characteristics with other machine learning algorithms: It starts with training data, it is supervised (the training data set has the outcomes), and it helps make predictions based on the data.

For example, say that you have a file that shows the cost to ship a truckload of product and the distance that the load traveled. Say that the sample data set has 296 observations. You want to understand the relationship between distance and cost. In this case, the cost is the dependent variable since the hypothesis is that it *depends* on the other variables, which are called independent variables. Therefore, the distance is the independent variable. By running a regression, you can then predict the cost of a truckload, based on the distance the truck will travel.

To break it down to the absolute basics, regression analysis works by finding the best possible fit for a line running through the data points. In more advanced regressions, the line does not have to be straight. In Figure 4.5, you can see the data plotted on a graph and the results of the regression—the line running through the points. This line can be expressed as a simple equation and then used for future predictions.

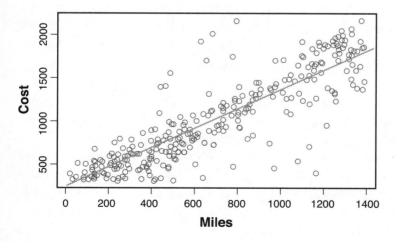

Figure 4.5 Regression Plot Example [5]

Regression is a powerful tool for making predictions and understanding relationships between the variables. It has widespread use in business, science, and the social sciences. In Chapter 6, "Predictive Analytics," we'll discuss one of its most popular uses in business: demand forecasting.

Since regression is so well covered in business schools, universities, and even high schools, we don't cover the details of the algorithm. (We recommend some useful resources at the end of the book.[6])

In this chapter, we do, however, point out some interesting facts about regression as it relates to machine learning. We think these facts will help give you a deeper understanding of machine learning in general.

First, it is interesting to note what you need to take from a machine learning algorithm when you apply it to new data points. In regression, no matter how large the training data set is, the result is a simple equation. So, for any new data point you would like to predict, you simply apply the equation. This is again in contrast to the k-nearest neighbor algorithm, which uses the whole training set every time. But it is similar to the decision tree, where the result is something you could create a program for.

Second, regression works on purely quantifiable data. In this section's example, you used truck cost and distance traveled, and the regression created an equation. You can plug any value for distance into the equation and get a prediction for the cost of the truckload.

You should also note that regression can convert some categorical data into binary variables that can only take on a value of 1 or 0. For example, if you wanted a variable for whether the truck was refrigerated, you could assign refrigerated a 1 and not a 0. These binary variables can only be used as independent variables in linear regression. These binary data points work because the regression equation is a set of coefficients multiplied by the independent variables and summed together. So a binary variable would simply turn on or off one of the coefficients. In this case, the coefficient for refrigerated or not may turn out to be $175. So, when the truck is refrigerated, you would multiply the $175 by 1, and this $175 would be added to the cost of the truck. When it is 0, no extra cost would be added.

Something with more than two categories would need to be quantified with multiple binary variables. For example, if a category like truck temperature can have three possible values (frozen, refrigerated, or normal), then you need to create a binary variable for frozen with a value of 1 or 0, and refrigerated with a 1 or 0, and then it would be assumed that normal would be the choice when the others are both 0. You can't just convert a single variable with values of 1, 2, and 3 because it would imply that being normal is three times as expensive as frozen and 50% larger than refrigerated. You would be assigning a numeric relationship between the variables.

Note that decision trees don't have this quantifiable requirement. In a decision tree, each column of data could simply be words or labels (like frozen or refrigerated). The algorithm treats everything spelled the same as being the same. And, in k-nearest neighbor, the outcome values could be text values or labels; in this chapter's example, you needed the other input to be numeric so you could measure distance, but the outcome variable doesn't have to be numeric. This is what makes the collection of machine learning algorithms so powerful: They don't all solely rely on quantitative data.

Third, regression analysis tests itself. In other machine learning algorithms, you hold back a random sample of the training data, which becomes test data, so you can test how well the algorithm performs. With regression analysis, you typically run with the full data set from the start. Then, various test results are reported back to you, and they will indicate how well the equation predicts the relationship between the variables. For example, you may remember from previous work with regression that the *R-squared* value tells you how well the equation explains the relationship between the independent and dependent variables. The *p-values* provide information on whether a specific independent variable has an impact that is statistically significant. If you've forgotten about regression, these terms may sound technical, but if you run a regression in Excel, you will see these same terms reported. In other machine learning algorithms, the math behind the algorithms is not as well developed, or it is impossible to get this more detailed resulting information. This is why you need a test data set.

Fourth, regression provides information around the confidence of the predictions. That is, not only does it make a prediction, but it provides a range for that prediction. For example, the regression may predict that a truck going 800 miles will cost $1,000. The regression will also tell you, with 95% confidence, that the truck will cost between $1,100 and $900, for example. Having this range can be very helpful in decision making. If the range were between $1,800 and $200, it would tell us that there is so much variability in the data that we should be cautious about using it as a basis for our decisions.

Fifth, regression is a well-developed discipline, and the analysis of regression results can tell a lot about data analysis in general. One example is called *omitted variable bias*. This is well highlighted by an example from the book *Managerial Statistics*.[7] In this book, the authors show that if you do a regression analysis using only strikeouts (a bad thing) to predict the salary of top Major League baseball players, you will find that the more a player strikes out, the better he is paid. This seems completely counterintuitive. You would think that a negative, like more strikeouts, would lead to a lower salary. The problem is that this example considers only one variable: strikeouts. It turns out that players who hit a lot of home runs (a good thing) also strikeout more. So when you do the analysis with just strikeouts, the strikeout variable is asked to carry hidden information about home runs as well. Omitting the variable home runs caused bias in the strikeout variable. If you include both home runs and strikeouts in the regression, you see that salary goes up with home runs, and it goes down with more strikeouts. When both variables are in the analysis, you can think of the regression as holding the home runs constant as it works out the proper equation for strikeouts.

A good general lesson from this example is that it is always important to have context and a general understanding of a problem. By understanding the baseball problem, you can see that something is strange in strikeouts leading to a higher salary. Knowing this caused you to revisit your model to gain a deeper understanding. This is why it is advisable on any analytics problem to solicit the guidance of those who have deep experience in the area being studied, even if those people will not be involved in the technical analysis. Sometimes the analysis will refute some things their experience tells them, but very often their experience will be invaluable in identifying important context to the analysis.

Standard regression tests can help you find other problems with the model, besides omitted variable bias. These include tests that show some variables are so closely related to each other that you can't separate their effects, to detecting outliers, and to detecting the fact that there may not be a linear relationship between your variables at all. In the case where there isn't a linear relationship between the variables, you can build nonlinear quadratic and log models—which may be complex but also make regression even more powerful and flexible in the long run.

Unsupervised Machine Learning

In unsupervised learning, there is no particular output or answer you are trying to find. Therefore, you are unable to test the algorithm against past data. In this case, you are running unsupervised algorithms to simply see what patterns emerge, hoping to gain new insights. In some sense, this is not as ideal as supervised learning because it can be harder to find information that you can take action on. In other cases, trying to uncover patterns from a set of data can provide great insight that would have never been discovered otherwise; you may not even realize what you were looking for when you started.

The two most common unsupervised machine learning approaches are clustering and association rules. Clustering is most often tied to the k-means algorithm, while market-basket analysis (what products are likely to be bought together) is a popular association rule.

Clustering and K-Means Algorithm

Clustering algorithms are probably the most popular form of unsupervised machine learning algorithms. In clustering and for other unsupervised algorithms, you are asking the algorithm to find patterns. You don't have a preconceived notion about what kind of answers you are looking for. Instead, you are simply looking for the patterns and the interesting insights they might bring.

The k-means algorithm is one of the most widely known algorithms in clustering and determining patterns in data.

With the k-means algorithm, you pick the number of clusters you want—again, this is what the *k* stands for—and then the algorithm breaks the data into that number of clusters, grouping the most similar items together in each.

The k-means algorithm is often used for demographic studies. If you have a set of data that describes people along various dimensions (where they live, their income, whether they travel a lot, how often they attend religious services, whether they play sports, the size of their family, and so on), you can then let the algorithm create natural groups. It is then up to the analyst to figure out what (if anything) the groups mean.

Several years ago, politicians who were analyzing groups of voters came up with the concept of the "soccer mom." This was most likely the result of a k-means clustering type of analysis. When the political analysts looked at this cluster, they likely found that it contained a lot of more married women with children, who lived in the suburbs, drove a minivan, and had several kids in school participating in many different activities (soccer possibly being one of the most popular ones). The k-means algorithm definitely didn't come back with a result that called this group "soccer moms." Instead, an analyst saw the general characteristics of the group and assigned it the clever name "soccer moms." The name likely stuck because it had a lot of truth to it. This newly discovered group of similar moms might not have been there in previous generations. The k-means algorithm brought it out in the open.

Once you've found such groups, the idea is to target your marketing, messaging, service offerings, and potentially products directly to them. For example, it became important for politicians to woo the soccer moms with special messages and for marketers to pitch particular products to this group.

Once groups are identified, you start to cross over to supervised learning. Once you know a group such as "soccer moms" exists, you can predict other women who fall into this group based on their characteristics. But you need to remember that you didn't know that such a group existed before you started the study.

You might be thinking that k-means clustering can be a bit cloudy in nature: You have to pick the k-values, and you then have to figure out what the groups mean. And, in that sense, it is cloudy. You have to do work to run the analysis with different values for k and spend time analyzing the output. In the study that found soccer moms, if the k value had been smaller, the people in the "soccer moms" group may have been absorbed into another group, and it might not have been obvious that this new group existed. If the k value had been larger, this group may have been split up into many other smaller groups and mixed in with different types of people (men, single women without kids, people who lived in the city, and so on) and would again have made it hard to pinpoint the group.

To understand this further, we should cover a little background on how the algorithm works.

The k-means algorithm works by picking the best k points that minimizing the system-wide total distance from each data observation to those points. Those points then represent the "average" value for all the data observations assigned to that point. The distance, as with the k-nearest neighbor algorithm, can be defined in various ways and may need to be normalized so you can work in multiple dimensions. For example, if two dimensions are income and number of people in the family, you would need to normalize these data elements so that income would not swamp the number of people. So, in our soccer mom example, one of the k points turned out to have many different women assigned to it. And these women assigned to it had the characteristics mentioned above. However, this k point may also have had some men and some single women (who shared characteristics with the soccer moms) assigned to it. The k-means algorithm doesn't necessarily create clean clusters or groups.

It turns out that finding the k best points that minimize the overall distance is actually an extremely hard problem. In fact, this problem is so hard (mathematically), that you can't actually prove that you've found *the best* k points. Luckily, there are approaches that help you find reasonable answers quickly. In a very insightful blog post, Carnegie Mellon Professor Michael Trick, in his popular Michael Trick's Operations Research Blog, pointed out that although the problem is hard in theory, the algorithms seem to work well when there are natural clusters.[8] That is, when your underlying data really has distinct clusters and presumably you hit upon the right value for k, the algorithm excels.

This is a good insight for managers using the k-means algorithm. If you pick a value for k, and the algorithm goes slow or doesn't show you much, it may be that there aren't k natural clusters. You should then try a different value of k. For example, if you are clustering people into three groups, you may find that the groups all seem to have equal numbers of men and women, equal numbers of people in different income brackets, and equal numbers of people who live in the city versus the suburbs. That is, there may be no real natural groups for the algorithm to find. By switching k from 3 to 6, you may see the data break up into much more natural groups. Or you have to be ready to accept the fact that the data may have no natural clusters.

Besides having to try many different values for k, you also have to analyze the groups to understand what they mean and whether the groups have value. It would be great if these groups could be automatically named and put to use. (We are sure someone is working on this as you read.) But, for now, it is up to you to analyze the different groups.

For example, if you analyze products sold in a store and want to group them based on their average weekly sales and their weekly sales variability (how much the sales fluctuate week to week and a good indicator of your ability to accurately forecast their sales), you can create a simple grid with each circle representing a product. It is also possible to visualize the results of the k-means cluster when you have only two dimensions. For example, Figure 4.6 shows what happens when you ask for three clusters; the dark lines show the boundaries.

In this case, the products break up nicely into three groups. The upper left shows products with high sales and low variability. For these products, you can safely order in large batches because you know they sell in high volume and are predictable. The lower right shows products with very low sales and high variability. For these products, you might decide not to stock them or to only order small quantities because you can't predict when they will sell. And the third group represents products with a medium amount of sales and variability. For these products, your strategy would be a mix of the other two strategies.

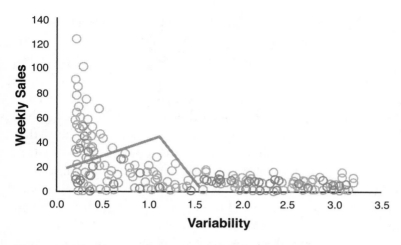

Figure 4.6 Sample Two-Dimensional K-Means Result

You could add more dimensions to the analysis—like the profit margin of the products or the amount of space the product needs at the store—to get additional insight into the product groups. (Of course, you would lose some of your ability to easily graph the different groups.) But, like before, it is now up to you to determine what you will do with these different clusters.

We'll come back to clustering later in the book—it has a wide range of applications. Here we touched on clustering customers (or voters) and clustering products in the store.

Association Rules and Market-Basket Algorithms

Association rules are another popular type of unsupervised learning. In this type of analysis, you are trying to see what items are purchased together, what events occur at the same time, or what characteristics are associated with each other. The most popular association rule is the market-basket analysis. This analysis gets its name from its most common use: determining what products are most often bought together in a store.

A market-basket analysis does something that sounds very simple: It looks at all the things that people buy together. There are some easy observations that this analysis can determine, such as people buying hotdogs and hotdog buns together. Of course, this is trivial, and you don't need a machine learning algorithm to figure it out. The algorithms are more valuable in finding associations you would have never thought of.

A popular example of this type of analysis is quite controversial: Some argue that the story is true and others that it is simply an urban legend. True or legend, it goes like this: One grocery store found that people commonly bought diapers and beer together. The explanation proffered in retrospect was that new fathers would run to the store to pick up diapers and also buy beer at the same time (although as both authors of this book are fathers of three, we don't recall having much time to enjoy beers while our kids were in diapers). As the story goes, the grocery store could then put beer and diapers near one another to further encourage this buying behavior. Although most of the evidence seems to suggest that this story is at least partly exaggerated, its colorfulness does help illustrate the potential power of these types of algorithms.[9]

There is no way a human being could possibly evaluate all possible associations that may exist within all grocery store purchases, but algorithms are great at doing this. The fact that a machine learning algorithm has made status as an urban legend is good for explaining what this field does as well.

Like the k-means algorithm, the math behind association algorithms is actually quite hard and the algorithms can take a long time to run. Frankly, the details of the algorithm don't help you build much intuition, but the general idea can help explain why it takes so long. Basically, the algorithm needs to look at every possible combination of items that can be purchased together. The total number of combinations in these data sets can quickly exceed the memory of even the largest computer. Think about a grocery store. There can be up to 40,000 different items on the shelves at one time. If you buy just 100 items, there are so many different combinations of 100 items (more than 10^{302}) that any computer would run out of memory if it had to consider all of them. Luckily, good algorithms exist to help you quickly sort through the analysis and find items that sell together faster.

What is most important is that the end result of these algorithms is a list of items that go together. And, since it is unsupervised, you need to analyze the results to see if you gain new insight from the results.

As with other machine learning algorithms, the line between supervised and unsupervised can blur. Association rules can also be run in a more supervised fashion. For example, you can pick a certain item, such as orange juice, and run the algorithm to see what other items tend to be bought with orange juice. This allows you to explore specific areas in more detail.[10]

A Note About Over- and Under-Fitting Your Models

If you read too much about the hype of machine learning algorithms, you will get the false impression that these models might be trivial to set up: Take a big data set, feed it to a machine learning algorithm, and get results. If only it were that easy. When building any of these models, you need to be very careful about setting up the models and fine-tuning the parameters—as you would with most any other analytics project.

When you are managing a project with machine learning, you specifically need to watch out for under- and over-fit models.

An under-fit model is usually easy for a manager to spot. An *under-fit model* is a model that does not use enough explanatory or input variables to predict the outcome. When you test it, it doesn't seem to be much better than a roll of the dice. And, when asking questions, you quickly realize that it is not actually using many inputs to drive the prediction at all. An under-fit model for predicting whether someone will buy a new car this year may only include a variable for income level. This factor might play a small part in predicting whether someone would buy a new car, but it would leave you asking about many more variables. For example, you might want to know the age of the buyer, the age of his or her existing cars, and the age of his or her kids (if any), among other things. Likely when you test the model with only income, you will find that the predications are not very good.

An over-fit model is much easier to slip past a manager. And, in some cases, managers may cause their analysts to create over-fit models. An *over-fit model* is a model that uses too many variables to explain the output. Preventing these models is difficult because they often do a great, or even perfect, job of predicting the results based on the training data. That is, when you run the algorithm and see how you did with the training data, you get results like 80%, 95%, or 99% prediction accuracy. This seems great. And, for an analyst, it may be fun to fine-tune the parameters and watch your predication accuracy increase on the training data. As a manager, you might be tempted to keep the pressure on your analyst to keep fine-tuning to go from 70% accuracy to 80% and higher. The problem with an over-fit model is that it may not predict new data points very well. Since it is over-fit, a lot of the input variables are just noise and actually cause the important variables to get lost. In addition, this type of model is not good for management decisions. It doesn't allow you to understand the key variables.

So what causes an over-fit model? The most obvious cause is that it has too many explanatory variables relative to the data set. For example, let's return to the model for predicting a new car buyer. If you have a training data set of 100 people, and you expand your list of explanatory variables to between 10 and 20 (with all the variables also able to take on a range of values), then a machine learning algorithm may be able to construct a set of rules so that each customer has its own unique parameters. So the algorithm, run against these 100 people, will predict their behavior perfectly.

Hopefully, you can now see the issues you need to watch for to prevent over-fit models. Let's now discuss how you can learn to avoid this in your models going forward.

The first issue is that the number of explanatory variables should usually be much smaller than your data set. This allows each unique combination of explanatory variables to have enough samples to make it meaningful. For example, if you had 500,000 people and the same 10 to 20 variables, you would have a good number of people in each of the buckets. With 100 people, you have buckets of 1 person; the algorithm essentially can then adjust for every nuance in the data, and the rules have no sense that some of the points may be outliers or not representative.

The second issue is that if you test against the training data, you can see outrageously good predictions. This is why you need to be careful to create both a training set of data and a test set of data. An over-fit model might capture all the nuances of the training data, but it is then likely to fail on the test data.

As with any other sophisticated project, there are many ways a machine-learning project can fail. It is good to keep this in mind. Just because machine learning is an exciting new area doesn't mean that your project is destined to succeed.[11]

Other Machine Learning Algorithms and Summary

There are many other valuable machine learning algorithms that we could have discussed in detail in this chapter.

For example, there are algorithms in the area of text recognition that go far beyond k-nearest neighbor for determining spam emails based on the types of words in an email. *The Economist* magazine reported on a technology for determining fraud by analyzing emails (to look not only for certain words but also for changes in behaviors), analyzing expense reports, and analyzing relationships.[12] This technology relies on machine learning to figure out the meaning of text. The company NarrativeScience, which was founded, in part, by people with a journalism and communications background, has reversed text-based analysis. It has created a product called Quill that takes in raw numeric data and outputs a written story. The idea is that people are much better at understanding a story than at looking at a spreadsheet or nice visualization. In their public talks, the NarrativeScience folks give an example where they can take a box score from a baseball game and write an article that is comparable to what is seen in the sports section. In fact, it can be hard to tell that their story was written by a machine. They have found a nice niche for this in Little League games by creating quality articles for kids and parents.

In this chapter, we talked about just a representative sample of these types of algorithms. Other authors may have picked a different and equally reasonable sample to highlight. Our objective was to give you a good flavor for the types of machine learning algorithms that are being used and, in many cases, are impacting your day-to-day life without you even knowing it. We also wanted to help educate you on the terminology and important issues in machine learning so that you would be in a better position to research or ask questions when you see a new machine learning algorithm.

You'll see these algorithms again when we talk about descriptive, predictive, and prescriptive analytics. We wanted to discuss them in their own chapter because they may be new to you if you are just learning about the field of analytics, and they are often closely linked to analytics in the general business press.

5

Descriptive Analytics

Descriptive analytics helps you understand what has happened or is happening in your business or organization. In the broadest sense, descriptive analytics allows you to make sense of the data you are collecting. Some would say that it allows you to turn your data into information and insight.

You may think this sounds a lot like just reporting on the data—and we agree with you. And, of course, reporting is nothing new. Organizations have always had to pull together monthly, quarterly, or annual reports about the business. Organizations have also always had to pull to together reports for the routine tasks of paying workers, paying commissions, tracking sales, or paying vendors. And, of course, many other reports have long been generated to help people better understand productivity, service, or other measures. There is probably some truth to the idea that by using the trendy name *descriptive analytics* you will get more attention and sales than if you had used the term *reporting system*. However, there are enough new things happening in this area to make people realize that good reporting is a class of analytics that can help make better decisions. It is also the case that the word *reporting* make many people think of tables with rows and columns of numbers. As we will see, descriptive analytics is meant to cover a much wider variety of methods used to gain understanding and insight from data. (In this book, we use the industry-standard term *descriptive analytics*, and you can decide if this word should be replaced by *reporting analytics*.).

There is a reason that of the three classes of analytics (descriptive, predictive, and prescriptive), descriptive is by far the largest: Organizations devote much more money and time to this class. The reason is simple: Organizations are very complex, and it is very difficult to get timely information to help make better decisions. And, although the authors of this book are biased toward the fancier math in predictive and prescriptive analytics, the truth is that many decisions are not very complicated. If managers can better understand their business by having the right information, it becomes easier (maybe not always easy) to make the right decisions.

For example, hospitals have long tracked important outcomes. One such measure was the number of patients who developed infections through bloodline catheters. These infections are very serious and could lead to the death of the patient. An article in the *New York Times* estimated that in 2006, about 80,000 ICU patients got such an infection, and 26,000 died as a result. The data and problem are well known. The solution was harder to find. But Dr. Peter J. Pronovost came up with the idea to use a simple checklist prior to inserting the catheter. The checklist included such basic items as washing hands, cleaning the patient's skin, wearing a sterile mask, and putting sterile dressing at the catheter site. It seems trivial and a bit shocking to an outsider that these steps weren't always followed, but in the busyness of an intensive care unit, simple steps such as these can easily be missed.[1]

To see if the checklist worked, researchers tested it in hospitals in Michigan for 18 months. The results were quite amazing and clear. By following the checklist, infections went to virtually zero; an estimated 1,500 lives were saved.

The descriptive analytics were simple in this case. The researchers had data on the number of infections before the test and during the test. They didn't need any complicated math to tell them how the test turned out. They simply looked at the data and saw that this was a success.

Although the bloodline example was extreme in that it involved life and death, it highlights that many correct decisions can be trivial, once you see the data. For example, if Machine A and B do exactly the same thing, but A is cheaper, you don't need any additional analysis to know to pick A. The problem has always been that without the proper presentation of data, we may never know this. (The skeptic may point out that even though the organization knows the best decision, it still doesn't do it. That may be true, but that is the subject for another book.)

So, descriptive analytics is about being able to describe data in a way that allows you to better understand the facts and make good decisions. What makes up good descriptive analytics? We will break it down into four major categories:

- **Getting information from and being able to explore your existing databases**—Most organizations already collect a lot of information. This information is or should be stored in databases. Managers need to understand how to access this information. In fact, we are seeing trends where non-technical MBA summer interns are required to know or learn how to access data in databases through SQL (covered later in the chapter). So managers can no longer rely on the technical IT staff to retrieve information. Managers will need to know how to do this themselves.

- **Good visualizations**—This is much more than just creating pretty graphs and charts. It is about conveying as much information as possible in a format that allows people to quickly understand the information being reported. This would include good graphs, charts, maps, and infographics.

- **Descriptive Statistics**—Although people naturally collect a lot of basic statistics (such as averages, minimums, maximums, counts, and so on) when they work with databases, it is also important to look at measures of variability and correlation.

- **Machine learning algorithms**—Good descriptive analytics requires the use of machine learning algorithms to uncover patterns in the data that you would never find with just reporting or by creating good visualizations.

We'll cover all four of these areas in more detail. We'll spend the most time on databases since this is the foundation for doing descriptive analytics.

Descriptive Analytics Through Databases

Descriptive analytics through databases is most commonly associated with business intelligence (BI) systems. Vendors that sell BI systems might not agree, but you can think of a BI system as a reporting system. But don't underestimate the value of a good reporting system. Remember that the vast majority of a company's IT infrastructure was built to run the business, run the cash registers, order products, pay bills, and prepare financial statements. These systems do not necessarily allow for easy reporting, descriptions of your business, or allow you to understand nuances within your business.

Without a BI system, it can be surprisingly hard to answer basic questions about your business (for example, "How many green shirts did we ship to Atlanta?"). A good BI system brings together data from your transaction systems and other systems (almost every firm has different types of systems) and links the data in ways that allow you to create standard and ad hoc reports to better understand your business. One of the keys to a good BI system is that the data from different sources is connected, so you have the flexibility to summarize the data or drill down into the details—without knowing, in advance, what reports you want.

Another way to think about a BI system is that it turns *data* into *information*, and it helps turn *information* into *knowledge*.

By *data*, we mean the raw facts about the company or organization. This could be information about customers, donors, suppliers, employees, sales, shipments, inventory, and so on. A raw fact would be the address of a customer, the fact that a customer bought a green shirt last Thursday, or that there are 10 green shirts in the warehouse. It is just a fact and nothing more.

By *information*, we mean that the data has been processed in some way to give additional insight. That is, if you say that you sold 500 green shirts in June, that information—we added up all the green shirts sold in the month of June—can allow you to take some action. If 500 green shirts is a lot, you can use this information to order more green shirts or remember to stock green shirts in June. You need to be able to process the data effectively to come up with the information. And if you process the data in new and creative ways, you come up with new information that may lead to changes in how you think about the business.

Knowledge is harder to define and can lead to long discussions beyond the scope of this book. But, for our purposes here, it refers to familiarity with a topic that can include understanding the data and information around that topic. It is usually earned through experience or study in the area. We only include it here to mention that vendors of BI systems would love to be able to automatically provide knowledge. Although some vendors and researchers are working on making that possible, just understanding how to get good information from your databases is our goal here.

In the rest of the chapter, we'll provide you with background on databases and business intelligence so you will be able to better navigate the different terms and applications.

Database Basics

Whether you are implementing a large-scale BI system or creating an Excel spreadsheet to help with descriptive analytics, you understand the need to organize your data. The early designers of databases realized this as well. In fact, a database comprises of data (facts) and structure (organization). The solution that quickly evolved and is still the standard today is the normalized relational database.

Understanding databases might seem like a function for the IT department and not for managers. However, as a manager, you may have to oversee a project or group that creates or maintains a large amount of data. This data will likely be stored in a database. If you have a basic understanding of databases, you will be in a better position to manage the project and make better decisions. In addition, we have found that managers who understand databases are in a much better position to effectively use the data they have. Understanding databases allow you to explore the data on your own and create reports and graphs. In other words, managers who can take advantage of databases will be better at descriptive analytics.

So a database is simply data stored with structure. The structure of the database provides integrity rules for the data type for each piece of data, the relationships between different data, and constraints on how the relationships work. We'll cover each of these in more detail in this section.

The heart of the database is the table. A table, like the one shown in Figure 5.1, comprises rows and columns. The rows represent the data that describes an instance of the entity. The columns represent the attributes of the entity.[2] In the example shown in Figure 5.1, each row represents a person—his or her name, gender, city, state, and language spoken.[3]

ID	Last Name	First Name	Gender	City	State	Spoken Language
1	Meyer	Connie	Female	New York	NY	English
2	Johnson	Olive	Female	Dallas	TX	English
3	Miller	Raymond	Male	Atlanta	GA	English
4	Clark	Clarence	Male	Chicago	IL	English

Figure 5.1 Sample Database Table

A single database table is a lot like how you use a worksheet in Excel. However, Excel can fail in many different ways. And, if you are a manager who has only worked in Excel, you may not realize the shortcomings. Once you see the power of databases, you may find that you convert your large spreadsheets to much more efficient databases and increase your descriptive analytics capabilities.

Excel falls short in the ability to enforce the data type of each piece of data. For example, the ID field should only be an integer, the name field only text, the gender only "Male" or "Female," and the language a list of choices. Now, Excel can do some of this, but making it happen requires some expertise beyond what most users have.

Where Excel really breaks down is in its inability to create relationships and constraints between data in separate tables. If you are used to storing data in Excel, you have likely seen this when you have data on multiple tabs and you have to link this data together.

A database really starts to get its power when you have multiple tables and relationships between the tables. Another way to think about this is that in many cases, different types of data should be stored in different tables because the different types describe different things. This is where a normalized database comes in.

A database has different tables for different pieces of data, and a normalized database (or simply a well-designed database) is set up so that each piece of data is stored in only one table. Then the tables are linked so you can know what data belongs to other data. It is probably best to highlight this with an example.

Let's stick with our example of a table of people, such as the one shown in Figure 5.1. And let's assume that this table represents patients at a hospital. Each of these patients can have multiple visits to the hospital, and each visit to the hospital may involve multiple procedures.

One way to store this information would be in a large table, as shown in Figure 5.2. This table violates the principles of relational database design. Notice that the name of the patient is repeated multiple times. Notice also that details about a single visit are repeated, and the cost for the visit is repeated. Since the cost for the visit is repeated, you cannot simply add up this column, or you will double (or more) count the costs. Besides creating redundant data and creating extra storage requirements, this also creates a chance for data errors (by spelling a patient's name incorrectly, for example) and a logistical problem for making changes (for example, changing a patient's name requires you to find and change every instance of the problem).

ID	Last Name	First Name	City	State	Spoken Language	Visit	Cost of Visit
1	Meyer	Connie	New York	NY	English	19-Aug-13	$325.00
1	Meyer	Connie	New York	NY	English	19-Aug-13	$325.00
1	Meyer	Connie	New York	NY	English	20-Aug-13	$450.00
1	Meyer	Connie	New York	NY	English	20-Aug-13	$450.00
2	Johnson	Olive	Dallas	TX	English	21-Aug-13	$100.00
3	Miller	Raymond	Atlanta	GA	English	22-Aug-13	$200.00
4	Clark	Clarence	Chicago	IL	English	23-Aug-13	$700.00
4	Clark	Clarence	Chicago	IL	English	23-Aug-13	$700.00

Figure 5.2 A Table That Does Not Follow Relational Database Standards

The solution to this problem is to normalize the data into multiple tables. Loosely defined, *normalized data* means that no data elements are repeated in a table. If you find yourself repeating data, you should pull it out into a different table. To normalize the data in Figure 5.2, you could create three tables: one for the patients, one for the visits, and one for the procedures. Figure 5.3 shows an example of this.

Patient Table

Patient ID	Last Name	First Name	Gender	City	State	Spoken Language
1	Meyer	Connie	Female	New York	NY	English
2	Johnson	Olive	Female	Dallas	TX	English
3	Miller	Raymond	Male	Atlanta	GA	English
4	Clark	Clarence	Male	Chicago	IL	English
5	Meyer	Connie	Female	New York	NY	English
6	Meyer	Connie	Female	New York	NY	English
7	Meyer	Connie	Female	New York	NY	English
8	Clark	Clarence	Male	Chicago	IL	English

Visit Table

Patient ID	Visit ID	Visit Date	Cost of Visit
1	1001	19-Aug-13	$325.00
1	1002	20-Aug-13	$450.00
2	1003	21-Aug-13	$100.00
3	1004	22-Aug-13	$200.00
4	1005	23-Aug-13	$700.00

Procedure Table

Visit ID	Procedure ID	Procedure	Cost of Procedure
1004	7501	Lab Tests	$200.00
1005	7502	Specialist Visit	$400.00
1001	7503	Doctor Visit	$200.00
1002	7504	Lab Tests	$200.00
1002	7505	Scan	$250.00
1005	7506	Lab Tests	$300.00
1001	7507	X-Ray	$125.00
1003	7508	Doctor Visit	$100.00

Figure 5.3 Example of Normalized Tables

Notice that the data is linked by the IDs of the table. The linking IDs are referred to as primary keys and foreign keys. The *primary key* is the identifying ID in the main table. When you use that ID in another table to link back to the main table, it is known as a *foreign key*. For example, in this case, the patient ID is the primary key in the patient table. When this key shows up in the visit table, we call it the foreign key, and we use this key to refer back to the patient table and retrieve information about the patient for that visit.

Note that the same patient ID can show up multiple times in the visit table. This also highlights another key part of the database structure: the relationship constraints. In this case, the relationship is specified such that one patient can have many different visits, but each visit can have only one patient associated with it. The relationship constraints help ensure the integrity of the database and keep the data clean.

The same relationship holds for procedures. Within each visit, the patient can have multiple procedures. And you could even imagine an integrity check that made sure the cost for the visit was equal to the cost for the individual procedures.

The "magic" of databases is that the linked IDs in the table allow you to easily move through the database to pull the information you need.

Also note that when you visualize the tables, as shown in Figure 5.3, it is a standard practice to show the tables with just their fields and the links between the primary and foreign keys. It is also common practice to show the relationship constraints by using a "1" for one and an "∞" for many (or infinity, in the terminology of Microsoft Access). This shows that a single patient can have many visits. And, within a single visit, a patient can have many procedures. You can see this diagram in Figure 5.4.

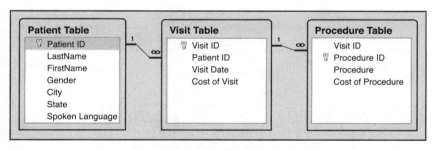

Figure 5.4 Relationship Diagram for Patients, Visits, and Procedures

In addition to normalizing the tables, you can also easily enforce integrity rules within a field. For example, you can specify whether the field is number or text, has to contain certain values (like "Male" or "Female"), or has to be in a certain format (like a date).

When you look at Figure 5.1, the original patient table, you notice that there is a field for the language spoken by the patient. It is very likely that the patients may speak multiple languages. If you have worked in Excel, you know that this leads to a problem called *overloading* a field—entering more than one piece of data into a field. For example, Figure 5.5 shows what is likely to happen if there is just one text field for the spoken language. Users will enter multiple languages in the same field. And if you allow such freeform text entry, you will also see misspellings and abbreviations creep into your database, which makes it harder to match data points. So you can see that overloading a field is a bad practice. In this case, it would get even worse if someone spoke three or more languages. Realizing this problem and if you only use Excel, you are tempted to create multiple language columns, such as Language1, Language2, and so on. But doing that is only a little better. The fields won't be overloaded, but now it will be more difficult to analyze the data for matches because sometimes "English" is in the first column and sometimes in the third column.

Using a normalized database structure solves this problem easily, as shown in Figure 5.6. With the languages spoken broken out into a separate table, now each person can easily have one or more languages spoken.

ID	Last Name	First Name	Gender	City	State	Spoken Language
1	Meyer	Connie	Female	New York	NY	English/Spanish
2	Johnson	Olive	Female	Dallas	TX	English
3	Miller	Raymond	Male	Atlanta	GA	French/English
4	Clark	Clarence	Male	Chicago	IL	English

Figure 5.5 Example with Overloaded Field

ID	Last Name	First Name	Gender	City	State
1	Meyer	Connie	Female	New York	NY
2	Johnson	Olive	Female	Dallas	TX
3	Miller	Raymond	Male	Atlanta	GA
4	Clark	Clarence	Male	Chicago	IL

Patient ID	Language ID	Language Spoken
1	101	English
1	102	Spanish
2	103	English
3	104	French
3	105	English
4	106	English

Figure 5.6 Normalized Tables with Language in a Separate Table

You can see the relationship between the two tables in Figure 5.7.

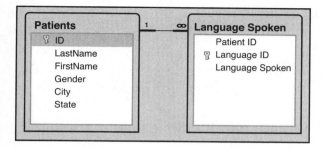

Figure 5.7 Relationship Between Patients and Languages Spoken

If you are using Excel, once you load your data into a database tool, the technology behind modern databases will open up a lot of possibilities for you. (Microsoft Access is a good place to start if you want to quickly get started on your own.)

A database gives you the ability to query the data with a standard language (SQL, covered later in this chapter). For small Excel spreadsheets, this is not noticeable. But when you have a large spreadsheet that covers several tabs, you quickly run into difficulties when you try to summarize the data in new ways.

For larger projects, database technology allows you to give people concurrent access. The business benefits of this are fairly obvious since many people often need access to the same data.

Finally, database technology allows you to scale your database. As you continue to increase the size of your Excel spreadsheet, it quickly bogs down and becomes unwieldy. Databases easily scale and allow very large data sets.

Data Modeling

Now that you've seen the power and flexibility of normalized relational databases, it is worth spending some time on how those databases get designed.

A data modeling expert designs a database much the way an architect designs a building. You wouldn't want to design a building without talking to an architect first. Likewise—and this comes as a surprise to many managers who are asked to lead a project that involves a lot of data—you don't want to create a database without getting help from a data modeler first.

So if you are running a large analytics project and need to manage data, it is important that you have a data modeler on the project. This section will help you understand what the data modeler does and why it is important.

There are three key steps in data modeling:

1. Creating the entity relationship diagram
2. Creating the relational model
3. Programming the database.

Let's take each in turn.

Creating the entity relationship diagram is the first step in the design of the database. The idea of this diagram came from a paper written by Peter Chen in 1976.[4] The purpose of an entity relationship diagram is to design a data model without getting into the complexities of a database; the diagram then acts as a communication tool for the data model. The communication aspect of the entity relationship diagram is a critical point for managers. It gives you a way for the technical team and the business users to work together to make sure requirements are captured. Without this tool, it can be difficult for these two groups to effectively communicate with each other.

Creating the entity relationship diagram involves defining the following:

- **Entities**—You should think of the entities as the real-world objects or people—things like patients, customers, products, or procedures. Often the analogy is that these are like nouns. They describe a person, place, or thing that you want to capture in the database.

- **Relationships**—The relationships define how the entities relate to each other. For example, a customer buys products or a patient receives procedures. The relationships are analogous to verbs (*receives* and *buys* in this case).

- **Attributes**—The attributes are the properties of the entity or relationships. For example, a patient has a name, gender, and languages spoken. Also, for example, a patient can receive multiple different procedures. The analogy here is that attributes are like adjectives or adverbs.

In the sample entity relationship diagram in Figure 5.8, rectangles represent the entities, diamonds represent the relationships, and ovals represent the attributes.

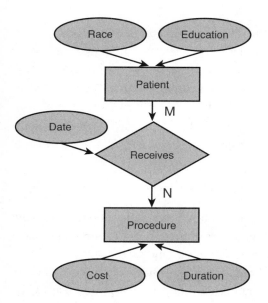

Figure 5.8 Sample Entity Relationship Diagram

The next step is to create a relational model. This is still just a logical model in the sense that you haven't written code for the database. But this logical model lists all the tables, the columns in each table, the primary and foreign keys in the table, and the relationships between tables. In other words, the relational model translates the entity relationship diagram into something similar to a blueprint for the database. This leads to the third step: the programmer uses the entity relationship diagram and relational model to program the database.

In some databases, like Microsoft Access, after you create the database, you can visualize the relationship model. You can see this in Figures 5.4 and 5.7. The line shows the relationship between the tables. It shows exactly which fields link the tables together. And the numbers at the ends of the lines show the cardinality of the relationship. In this case, a person from the People table can have only one gender, but a particular gender can be shared by many people. (Note that the one-to-many relationship is probably the most common relationship constraint, but you can also have one-to-one and many-to-many relationships.)

Whole books, courses, and degrees focus on database design and concepts. This section is meant to give you, as a manager, an overview of the basic principles so you can understand how to get started with this type of activity and help determine the questions to ask. It is also meant to help you realize some steps you need to take if you are confronted with managing a large analytics project that requires a complex database.

Learn SQL, Understand Your Data Better

Once your data is loaded into a standard relational database, you access the data by using a standard computer programming language called SQL (Structured Query Language). One nice thing about SQL is that it is an established standard for querying relational database software. All the relational databases that you encounter use SQL. You should be aware that there is nothing to force the different vendors to perfectly follow the standard, so there will be slight differences between how the vendors implement it. But no matter what software you use as your database, the basics of SQL and general concepts of SQL will be the same. Another nice thing is that SQL is relatively easy to learn. Also, some packages, like Microsoft Access, allow you to visually write SQL code by dragging and dropping the fields you want in your report.

We mention these pluses of SQL because many managers are intimidated by the thought of learning a programming language. Instead, think of SQL as a series of commands, not necessarily a full program.

We have found that business managers who know how to analyze data in Excel but don't know databases (or have never used Access) are extremely productive when they discover the power of SQL. That is, they can quickly create reports and analyze large data sets in ways that are impossible (or difficult) with Excel. In fact, one way to quickly get started with SQL is to load your big Excel file into a table in Access and start using SQL. If you have just one big table loaded into Access, you can start to use SQL. You don't need to even worry about normalizing the data. (If you want to load multiple tables and link them, you might have to worry about normalizing your data.)

One of the key trends with the proliferation of data is to make the data available to more people in the organization so they can better do their jobs, create the reports they need, and use data to do ad hoc analysis to answer questions as they arise. When you are doing this, you are putting the power of SQL into the hands of more people. Now, as in Microsoft Access, the users may not know they are writing SQL to do ad hoc analysis, but they are. Often, when people talk about gaining new insight into the business, they are talking about the ability to create ad hoc reports to answer specific questions. These questions may not be known ahead of time, so the reports must be flexible. In addition to flexibility, people talk about the ease of use of the reporting—that is, people who have a business question can create a report themselves.

All this means that descriptive analytics can be used deep in the organization. For example, if someone thinks a certain type of customer orders a unique combination of products, he can check it out. In the example of Circle of Friends we mentioned in Chapter 1, "What Is Managerial Analytics?" the CEO of the company most likely used SQL to look through his data to discover that it was moms who most frequently used the site. It was this insight that led him to change the company to Circle of Moms.

So what does the SQL language look like?

A quick internet search on "Basic SQL Commands" will return plenty of sites to get you started. There are commands that allow you to create tables, delete tables, add records, remove records, and so on. But, as an analytics manager interested in descriptive analytics, the SELECT command is what you will be most interested in.

The SELECT command allows you to pull fields from a table or multiple tables. (The FROM command goes with it, to specify the table you are pulling from.) You can then add other commands to this to narrow your search. You can use the WHERE command to create filters (for example, just show data WHERE the state is Florida or WHERE sales are greater than 100 units).

You can use the GROUP BY command to group together all record in a field that have the same value. For example, you can use GROUP BY to create a report that groups all the customers by their state or postal code. You can use the GROUP BY function for multiple fields in the same report. For example, if you group all customers by state and gender, you'll have data for all male customers in Florida and all female customers in Florida. If you group by state, gender, and age group, you'll have multiple age group data for all the female customers in Florida.

When you use GROUP BY, you will want to use standard mathematical functions on other columns—like SUM, AVERAGE, COUNT, MIN, MAX, and so on—to do things like sum the total bought by all the female Florida customers, or to count the number of unique customers, or to average their sales.

Part of the power of databases is the ability to report on data that sits in different tables. You can easily use the commands we discussed above on multiple tables after you tell SQL how the tables are related. The details of specifying that tables are related get a bit technical, but the principle is exactly the same as we discussed when building relational databases. Microsoft Access allows you to do this visually by allowing you to put multiple tables on the workspace and draw lines between the primary and foreign keys of the different tables. In SQL, the JOIN command handles this. There are multiple variations of the JOIN command that allow you to build different rules.

In a widely dispersed system, the work of joining relevant tables together should be of the realm of the technical team. But a manager should know about it. When you are doing reports on your own, you might need to join tables; you need to be aware that you can introduce errors like double counting or ignoring data if your joins are not correct. Also, we've seen some reporting tools that allow you to create nice graphs and charts that may require something like joining tables to correctly create the reports you want. The more you know about this (and you can learn the basics quickly), the better you will be able to do descriptive analytics.

Although this section has been somewhat technical, you should realize that you can make the power of SQL available to more business users with forms that sit on top of the database and hide the language of SQL. Microsoft Access allows you to easily create these forms, which may link multiple tables of data together and allow users to sort, filter, and view the data in a variety of ways. SQL is powering this exploration of the data, but the user is just interacting with a simple form and doesn't know anything about the SQL. Giving people these forms to interact with the data is a powerful part of descriptive analytics.

What Is NO-SQL?

Much of the technical hype around Big Data comes from the fact that some data sets are so large and unstructured that the traditional relational databases that we've talked about to this point no longer hold. For example, Facebook cannot manage all its data in traditional relational databases. It would not be possible for Facebook to keep up with all the picture uploads, likes, and comments and still present them in a timely manner on all the correct newsfeeds with traditional relational databases.

Companies like Facebook, Amazon, Google, and Twitter needed another solution, and a nontraditional database structure evolved. Google made this structure popular. Without getting bogged down in the details, Google needed a way to index everything on the internet and quickly retrieve the right information. It did this with MapReduce. The basic idea is that MapReduce allowed Google to process a job in parallel on many different computers. Systems like the open source Hadoop incorporate MapReduce and other technologies to allow other companies to take advantage of parallel storage and processing of massive unstructured data sets. Hadoop is just one of many systems that do this.

The term NO-SQL has come to describe these systems. The term NO-SQL evolved to mean "Not Only" SQL—as opposed to meaning no SQL needed. Presumably the early hype that SQL was outdated and no longer needed was a bit premature. SQL and relational databases are very powerful, and even very large data sets can take advantage of this proven technology. So don't let anyone convince you that SQL and relational databases are outdated. They should still be part of your descriptive analytics toolkit.

Unlike with SQL, there is not an established standard for NO-SQL. Each product has a different language and a different way to organize the data. However, the basic idea of a NO-SQL database is that data is not stored in a table but instead in a simple file (think of a page) with tags. The file does not have to be structured and defined ahead of time. For example, in a relational database, you need to predefine the field for gender in the patient file. In a NO-SQL database, you can simply add this tag later.

A NO-SQL database is also distributed. This means that each file sits on multiple servers. When you make a change to one version of the file, the changes are eventually propagated to other servers to update the file on other servers.

As a simplified example, think about the fact that Facebook stores a file about you on each of its servers around the world. When you make a change on your Facebook page, Facebook first updates your file on the server you are logged in to (the server to which Facebook directs you, based on where you are). Then the Facebook server sends out messages to all the other servers in its system to also update your file. But this is not all happening instantaneously. For example, if you are in the United States and your friend in India looks at your Facebook Wall immediately after your change, she might not see the change because the local server in India has not yet updated your file.

This example points out a few interesting things about NO-SQL. First, all the data is not immediately in sync. This is okay for Facebook because your likes do not have to be updated all at once. Contrast this to your banking system. If you withdraw $200 from your account, this fact should be known instantaneously anywhere in the world. There can't be any latency; if there were latency, can you imagine the fraudulent schemes and violations of banking rules? NO-SQL gives you the ability to do massively parallel transactions. Since files are duplicated on many different servers, you can analyze and work on these files on different servers. This also provides you with a lot of redundancy. If one server goes down, you have the right information in many other servers and can quickly recover (maybe without the user noticing).

This same parallel storage capability also holds for running queries. A query, if possible, can be parsed into many different smaller jobs and sent to multiple machines. The results from these smaller jobs can then be summarized in the end. Interestingly, as the jobs are distributed, the central processing keeps track of the jobs, and if a particular machine does not return the results from its job, that job is reassigned. This allows for redundancy in processing as well. And it allows you to quickly process large jobs. Not all queries lend themselves to parallel processing, but those that do can take advantage of this technology.

These NO-SQL databases can handle massive and unstructured data sets. But they do mean you give up some of the nice features of SQL, like GROUP BY, JOINS, and easy integration with other systems that support SQL.

As a manager, it is good to know about NO-SQL. However, for most managers in most cases, relational databases with SQL will be the right way to store and analyze data. NO-SQL is better for the largest of the large data sets. If your data set is that large, you should have an expert in NO-SQL on your team to help out.

A Note on Structured Versus Unstructured Data

For all categories of analytics—descriptive, prescriptive, and predictive—you should be aware that the underlying data can be structured (for example, a standard set of numbers in a predefined format—such as data in a relational database) or unstructured (for example, a stream of social media feeds, a stream of video feeds, a collection of medical journals).

As a manager, you will mostly deal with structured data. And, most analytics techniques require structured data. For example, if you want to simply explore your data with SQL, the data needs to be structured. If you want to run some basic statistics or a machine learning algorithm, you need structured data. And the list goes on. In fact, most of the algorithms that you will use directly or that your team of data analyst will use will require structured data.

However, most new data is unstructured—such as videos, pictures, and social media feeds. The amount of unstructured data far exceeds the amount of structured data, and some exciting work in analytics is being done on unstructured data.

For example, IBM's *Jeopardy!*-playing Watson computer was built to analyze unstructured data, such as a collection of medical journals and patient histories. Shortly before she became CEO of IBM, Ginni Rometty spoke at Northwestern University for IBM's 100-year celebration. She mentioned that the Watson computer represented a new breakthrough in computing technology. In fact, she went on to claim that this technology represented just the third era of computing. In the first era, which lasted up until about 1960, computers were basically big calculators. The second era, which we are still in, is the era of programmable computers. Of course, this opened up a lot of power and allows us to create many incredible programs—think about all the programs you interact with on your PC, tablet, or smart phone. The third era, which is just starting, is the cognitive era. This is where computers like Watson can sort through unstructured data, form hypotheses, determine the best answer, and learn from mistakes. Watson is running very sophisticated machine learning algorithms. The Watson computer started with *Jeopardy!* but has quickly moved on to healthcare, where IBM thinks it can help doctors better diagnose and treat patients.

Time will tell whether these new techniques truly represent a new era. But what is not debatable is that this type of analysis is quite cutting edge and is likely to uncover many new insights and changes.

Also, with unstructured data analysis, you will hear a lot about the work going on with video analysis. For example, cameras set up to watch for crime or monitor automatic production need to be analyzed. Or, if a researcher wants to better understand consumers, he may want to analyze all the videos uploaded to YouTube. Videos are unstructured data, and they generate large data sets very quickly.

But keep in mind that a lot of what researchers are doing with unstructured data is figuring out how they can cleverly provide some structure to it so it can be analyzed. This will not likely be a traditional relational database. But a clever breakthrough may create a structure that is equally powerful for this new unstructured data.

To keep all this in perspective, coming up with ways to analyze unstructured data and creating "cognitive" machines is only one part of analytics. It may be a part of analytics that gets a lot of press, and it may be true that the amount of unstructured data swamps the structured data, but in the end, the analytics applied to structured data will still be extremely important to most managers.

To put this another way, just because one area of analytics is getting a lot of attention from researchers does not mean that other areas are not important. In fact, if we had to guess, we would say that most readers of this book will be involved with analytics using structured data.

Data Warehouses and Data Marts

We've previously discussed the importance of normalized relational databases and data modeling. So if you are managing a project involving data, you now understand why you need to think about the data model and that you are likely to create a relational database.

Also, you know the SQL is the tool that helps you create reports and explore data within databases. In fact, if you have access to data in Excel, you can simply load it into Microsoft Access and start using SQL.

Now that you understand the basics, we are going to cover how larger corporations and organizations provide their employees with the ability to run descriptive analytics on the data they have, using BI systems and the data warehouses at their center.

Why would large corporations and organizations need anything more than the ability to run SQL on the data they have? And wouldn't their data already be stored in relational databases?

Here is the apparent contradiction that drives managers mad: Most companies computerized their operations long ago (measured in decades), have teams of capable data modelers, and have collected massive amounts of data. Yet it is typically quite difficult or impossible (for technical reasons) for managers and employees to get their hands on this data to analyze it.

But the contradiction goes away once you realize what the systems were set up to do. These computerized systems were set up to run the business, not for descriptive analytics. That is, these systems were set up to keep track of inventory, record sales, pay vendors, manage employees, keep track of accounting to report earnings and pay taxes, and so on. For example, when you make an airline reservation, you are interacting with the airline's reservation system. When you purchase something at a grocery store, you are interacting with the store's sales system and possibly its inventory system. We can refer to these systems as *transaction systems* (which keep track of the day-to-day transactions of a firm).

It is clear that these transaction systems collect a lot of data that would be valuable for descriptive analytics. What is also very clear is that the team responsible for maintaining the transaction systems absolutely does not want anyone accessing these systems directly to do descriptive analytics. They weren't built for that, the team doesn't want you accidently corrupting the data (like erasing someone's flight information), and the team doesn't want you to run a query that slows down the system. If your customers can't make reservations because you are analyzing data, you'll have big problems.

This is the reason that BI systems were created. This is the reason that you hear terms like *data warehouse* and *data mart*. The idea is that the systems that run your business are left alone. But no one doubts that the data in those systems is valuable. So data is extracted periodically from those systems and stored somewhere else—like in a data warehouse or data mart (a smaller version of a data warehouse). You can then do your reporting or descriptive analytics from there.

We have oversimplified this process by stressing that using a data warehouse allows you to access the data in your transaction systems. It actually does more than this. If you were to be granted direct access to these systems, you might find it difficult to do descriptive analytics. First, since the system wasn't built for descriptive analytics, the data may be difficult to analyze (for example, customer ID fields may be used in different ways for different purposes), data may not be kept (for example, information on flights from more than three weeks ago may not be kept), or you might not find the data you need (for example, information about the product purchased may sit in different systems).

The other important job a data warehouse does is to put the data in a format that is useful for analysis. This involves many different steps. It often involves creating a fresh data model that creates the right relational model. For example, since some data in transaction systems is erased when it changes, the new data model may have to add fields so you can time stamp your data. It often involves pulling data from other sources in the organization. That is, your data warehouse may need to pull data from the sales system, the inventory system, the accounting system, and the online system. It often means reconciling data stored in different formats but represents the same thing. For example, if your organization is international, when you pull together data from different countries, you may need to convert to a single format for dates and money. You may need to watch out for simple descriptive fields that are different in different languages. Even if your business is not international, the same product may have different IDs in different systems. You need to map these together so you can understand them. Also, different systems may store data in different ranges. For example, if one system describes customers between the ages of 25 and 35 and another system describes customers between the ages of 20 and 30, you may be have difficulty coming up with a good range to use in the data warehouse.

So, you can see that creating a data warehouse is not trivial. It is worth discussing some of the key ideas for data warehouses so you are aware of them as you interact with your own data warehouses or need to lead a project that is creating a data warehouse.

Bill Inmon is recognized as one of the leading pioneers in the use of data warehouses. He has a four-part definition of a data warehouse:

- **Subject oriented**—Data is stored based on subjects (like sales, products, or donations). The transaction systems where this data is pulled may not be organized like this. So this gives you a basis for not simply replicating the data in the transaction systems.

- **Integrated**—A data warehouse will pull data from multiple transaction systems. As we discussed earlier, the transaction systems may not be naturally linked. Integration gives you the ability to do analysis across different parts of the business or organization.

- **Nonvolatile**—Data is not deleted from the data warehouse. The transaction system may delete data, but the data warehouse is meant to keep it so you can later go back and analyze that data. For example, if you notice some trend today, you might want to look back at last year's data and data from the year before and see if the same trend presented itself. If it did, you may have a found a rule that you can use next year.

- **Time variant**—Data is time stamped in the data warehouse. Data in transaction systems may not be time stamped and may be erased as data is updated. The data warehouse is meant to be able to "re-create" the past by using the time stamps. This is often critical when you're doing analysis because you often want to know the state of the business when certain events occurred and how the business changed after those events.

From the preceding discussion and definition, you can understand why it takes time to create a data warehouse and, therefore, why it takes time to get a full BI system up and running.

You can also see that you need to design your data warehouse so you can get the information you need out of it. When you design a data warehouse, you are deciding what data goes into it and how it will relate to other data. As with all other engineering design, you must make trade-offs. There are some high-level trade-offs between total cost to implement, flexibility, and time to implement. (And remember the old engineering adage: "You can pick any two of those.") Once you make a design decision, you are potentially eliminating the ability to generate certain types of reports or do certain types of descriptive analytics. So design is important.

There are two general data warehouse design principles. We describe them only at the highest level so you have a starting point for understanding the issues. But each approach has many nuances that you need to understand if you are part of such a project.

Bill Inmon advocated for a top-down approach. The basic idea of this approach is that the organization designs a central data warehouse for all the data. This approach attempts to pull all the data together for the organization in one place. Then, if different departments need a smaller data mart to analyze just information for their part of the business, the data mart can be extracted from the data warehouse.

If you research data warehouses, you will find that Ralph Kimball's name is right up there with Bill Inmon's in the ranks of early leaders in the area of data warehouses. Ralph Kimball advocates a bottom-up approach. In his approach, each group or department develops data marts for its part of the business. Then these data marts are linked together to form the data warehouse. So there is no overall design of the data warehouse; it is just a collection of the individual data marts.

Of course, there is no correct answer to the question, "What is the right approach?" Both have advantages and disadvantages. The top-down approach can ensure a comprehensive design, but it may take much longer to create and is likely less flexible. The bottom-up approach can be faster, but it could miss important data or miss opportunities to reduce redundancy. And, surely, each approach does its best to minimize its weaknesses.

Once you have data warehouses up and running, you have a platform for running descriptive analytics. With a data warehouse, you can unlock the potential of SQL within an organization by giving many people access to the data warehouse and letting them answer questions and uncover important trends.

The data warehouse is also the basis for a BI system. But, if a BI system were only a data warehouse, *BI system* likely wouldn't have caught on as its own term. Once the data warehouse is in place, you can do a lot more—which is why we describe this as business intelligence.

Dashboards and Balanced Scorecards: Good Timely Reports

When you have a data warehouse in place, you should be able to create standard reports that people can use to understand and monitor their business. That is, when that the transactional data has been put into a format that allows you to easily create SQL queries, you can create standard reports that help people in the organization make better decisions. Often, the data warehouse is designed so that it can generate these reports that people need.

Two special kinds of standard reports that can now be generated and are part of the value of a full business intelligence system (above and beyond a data warehouse) are dashboards and balanced scorecards.

Before the days of data warehouses and business intelligence systems, it may have been difficult to get timely information. Reports were run once a week or once a month to minimize the interference with the transaction systems or because it took some effort to compile data from multiple sources.

One of the important uses of descriptive analytics is to give people a quick view of their business. There is no need to have to wait for this information. These types of views or reports are referred to as *dashboards*. The reference is to what an airplane pilot sees—all the critical parameters and controls. You can use a dashboard to see at a glance what areas are performing well and which ones are not. You will often see dashboards use green, yellow, and red to signal potential problems.

Dashboards are not a new idea. But when used with a good business intelligence system, they make a type of descriptive analytics possible. And you can push dashboards deep into your organization— the CEO can have a dashboard, the sales team can have a dashboard, the production team can have their dashboard, the area manager can have a dashboard, and so on.

One area where you see dashboards is in online businesses. When your entire business is online, you can use dashboards to see how many people are visiting your site, what pages they view, and what they buy. So, with a quick glimpse, you can see your entire business. You can then quickly look at the dashboard for the last year, last month, or last day. This type of dashboard allows you to do detailed descriptive analytics on your business and look for trends and insight. Google has made this fairly popular with its Google Analytics product. If you have a website (or even a simple blog), you can install Google Analytics, and you will be able to see a dashboard of your website.

A dashboard does not have to be a fixed report. A good dashboard report allows a user to filter, sort, and show the data in many different ways.

Dashboards can become powerful management tools that allow you to better understand and react to what is happening in the business. However, if people are focusing only on the items that show up on a dashboard, you better make sure all the key measures are included in the dashboard.

This issue leads to the idea of using a scorecard or balanced scorecard to complement the dashboard. A scorecard tracks the information that a manager will be judged on. The idea is that there are a handful of measures that matter to the overall health the business, and you want to track them and measure your performance against them.

For example, a sales manager for an online retailer may have a scorecard that includes measures for the percentage of visitors who convert to sales and the average sales price per transaction. This person is expected to design the site so that when visitors come, they purchase and so when they purchase, they spend as much as possible. The marketing manager's scorecard may include the total number of visitors to the site. That person is responsible for getting the visitors to the site. And the CEO's scorecard may include measures for revenue and profit.

The "balanced" part of a scorecard is also important. It is a reminder that you need to make sure the scorecard includes both sides of key trade-offs. For example, revenue may be important. But if you just track revenue, you might lose track of the profit margin or the costs of sales. A good balanced scorecard makes sure that you can't artificially make one measure look good without hurting another measure.

The idea is that you use a dashboard to help control your business and make decisions. And when you make decisions, you are guided to make decisions that help improve the scorecard measures. And by doing that, if you've set up the scorecard correctly, you are making the business healthier.

To distinguish the difference between a scorecard and a dashboard, let's return to the sales manager for the online retailer. Her scorecard may contain percentage of visitors who buy and revenue per transaction. Her dashboard may contain a lot more information. For example, it may contain the number of pages a person visits, which pages get the most visits, and how long a person stays on the site. These measures are not part of the scorecard because if the sales manager could artificially raise the number of pages a person visits, it wouldn't necessarily lead to more sales. However, it is part of the dashboard because she knows that the number of pages visited impacts sales. Therefore, she tracks this on the dashboard and tweaks the content and design of the pages to impact this measure, with the goal of improving the measures on the scorecard.

One complaint about scorecards and dashboards it that they just tell you how you *are* doing but do not give you a measure of how you *should* be doing. For example, these tools might tell you that you are converting 5% of visitors to sales, but how do you know if that is a good number? This is where benchmarking comes in. You can get benchmark data in a variety of ways. The most common is that firms can either talk to other firms in a similar industry or use industry data to determine how other firms (typically the best) are doing on key measures. These numbers then become benchmarks for performance. So if the industry leaders are converting 10% of visitors to sales, your 5% does not seem so good. But if the best in the industry are converting 2%, you are doing great.

Sometimes it can be difficult to get relevant benchmark data from other firms; it may be confidential information, or it may be that the details of the business don't match up. In these cases, it becomes important to develop internal benchmarks. The easiest way to do this is to compare similar units. For example, you could compare sales per square foot of your different retail outlets.

Expanding the Reach of Descriptive Analytics with OLAP and Data Cubes

In this chapter, we have discussed the value of creating standard reports, dashboards, and scorecards. When you design reports, you can add a lot of power to them by allowing different filters, allowing different views, and even selecting different time ranges. These types of reports are a powerful part of a business intelligence system.

But with a data warehouse, you could generate an unlimited number of reports. This is why you give employees access to the data warehouse—so they can explore the data on their own, answer new questions, and uncover new insights.

You quickly realize that a data warehouse has a lot of data in it. This can mean that it takes time (like hours or days) to run certain queries. Also, it can mean that a normal business user can get lost in the data that is available. Data marts partially address this by allowing a group to have access to just the data that is relevant to them. But data marts also contain a lot of data. For example, think about all the sales data in a sales data mart: It may contain data from the past five years of sales, by product. This is still a huge database.

Also, different people in the organization need to be able to do ad hoc analysis with different sets of data. Some people need to look at the raw data, while others needed aggregated data. The solution to this problem from the business intelligence community was the creation of data cubes and OLAP (Online Analytical Processing). These terms may seem technical, but the key idea is that many or most people in an organization do not need to access every part of a data warehouse, and what they do need is a high-level summary of the data in the data warehouse. That is, most people don't need all the details in the data warehouse, but they do need to access the information quickly. One solution is to design a common set of data that people might want to analyze. So, in a sense, it is like designing a data warehouse. However, you just define the data by the key dimensions and how you want to summarize the data.

This will be easier to understand with an example. Let's say you work for a retailer that has many stores, and you want to analyze sales at the stores. You have a huge data warehouse with every sales transaction, what was bought together, and the time and cash register on which the sale was made on. This is too much information for most analysis. So you define a set of data that will be for every store, every week, and every product. For this set of dimensions, you want to know the total transactions and total sales. This new set of data is a *data cube*. (So we have data warehouses with the most data, data marts with less data, and a data cube with even less.) To create this data cube, you run a query against the data warehouse and determine the total sales and total number of transactions for every combination of stores, products, and week. And once this happens, the data will be stored in memory (this is the "online" part of OLAP) so you can quickly access this data.

Now you can quickly run ad hoc reports on the data: Which products are selling in which stores? What products tend to show seasonal trends? And so on. But you lose some of the detail in this process. In this example, you can't determine sales on a Saturday. You also can't see what items were purchased together in the same cart. This leads us to why this is called a *cube*. To stick with our earlier example, when we plot our three dimensions, we have an axis for store, product, and week number. This forms a three-dimensional cube. Think about this larger cube being made up of many tiny cubes. Each of the tiny cubes represents a unique combination of a store, a product, and a week. And, within this cube, you are storing information on the total sales and total transactions. That is, within the cube, you are aggregating a lot of data. You may have thousands of transactions that result in the two pieces of information that you are keeping. This is what makes cubes fast. And if you create them correctly, you have the appropriate level of information that allows users to still do ad hoc analysis without wrestling with the entire data warehouse.

It may seem like we are covering a lot of different ways to store and retrieve data. But it is actually hard to describe a complex organization in one uniform way. Business intelligence systems add features based on real needs. Different people need to see data at different levels. For example, the CEO needs to see one set of summary data that covers the whole business. This view may have one number representing the manufacturing plant's performance. The manufacturing plant manager needs another set of data that has a lot more details on the plant but no other information. And the European sales manager needs another view of the data. All the data just mentioned may come from the data warehouse, but there is no need for the different groups to have to interact with the data warehouse. So OLAP is a good solution in this sense.

Also, without OLAP, a lot of nontechnical people were not going to be able to run ad hoc queries. Giving them access to the data warehouse was going to lead to more problems than solutions. OLAP is a nice solution to give more people the ability to do ad hoc analysis.

When to Break the Rules of Relational Databases

Now that we've covered how to set up a proper relational database, it is probably wise to talk about when and why you should sometimes break the rules. One of the key rules we discussed is normalizing data.

When you run queries against a relational database, you often need to pull information from multiple tables. These tables need to be linked together, and the links are called *joins* in SQL terminology. For large data sets, the joins can slow down the processing time. The processing time can easily grow into minutes and hours. This is often too slow for the users of the data.

The OLAP and cube technology we discussed in the previous section were set up to help get around large data sets and having to wait for queries with many different joins to run. The data storage method of data cubes is often referred to as *dimensional* (rather than *relational*): The data is stored in a multidimensional cube.

Although OLAP can speed up some cases, sometimes you need fast analysis on the full set of data. In these cases, the often surprising answer is that you need to create a de-normalized single table with all the data. Sometimes this is referred to as a *flat table* (in contrast with cubes and relational databases). A query can typically run very quickly on a flat, but very large, table. For example, this can come up when you need to be able to query information about a customer when you are on the phone with that customer.

Besides speed, there are other reasons you might want to work with flat tables. It can sometimes be helpful when feeding data into an algorithm. The algorithm will not necessarily understand your specific relational database; instead, it will take a single table as an input. And it can sometimes be helpful to create non-normalized or flat tables when generating reports. Everyone likes to see reports in different formats and layouts. This often creates a need to break the rules of relational databases to create the reports. So, as a manager, as you design the layout of a report, you should keep in mind the programming required to get that layout. You may face a trade-off between an easy-to-generate report in a less-than-ideal format and a hard-to-generate report in a better format. It isn't just the report designer being difficult.

Real-Time Data and Automatic Alerts

As companies and organizations started to see the value in more timely reports and more up-to-date dashboards, the desire for more real-time data grew. They saw a nice benefit in going from monthly reports to daily or hourly dashboards, so why not push it further and make more and more information available in real time?

Combine the value of dashboards with the dramatic increase in the real-time data we are seeing, and you can understand how management wants more real-time analytics. We have more sensors on equipment that can send data every second (or more often). We may even be using video to watch parts of a business or factory (beyond its traditional use for security. This type of data is often referred to as *stream* data—it flows in fairly constantly at a high volume. You can contrast this to traditional data that arrives to the organization as a transaction—a sale or donation, a new patient, a shipment from a vendor, and so on.

This stream data is often handled by different types of technology and treated differently from a business point of view. It is not clear that you even want to store all this data. These data streams may only have value in real time, and once the moment is passed, the data is no longer needed. You may need different technologies to handle this data. If the data is arriving as a stream and you need to take immediate action, you might not have time to store the data in a traditional database and then run analysis on it. You might need to run your analysis in real time as well.

For example, with real-time data, you can set up algorithms to monitor the data and send alerts when there is a problem or trigger an action when certain conditions are met. Certainly this can be powerful. With dashboards, someone has to look at the data and decide to take action. Now, if you could program the knowledge of management or the rules management goes through to make a decision, and marry this with real-time data, you could automate decisions in real time. And if your algorithms were running predictive and prescriptive analytics, you could make even better decisions.

Before we get too far, we should define *real time*. If you talk to some managers or vendors of solutions, they will refer to *"real time"* as data that is available daily or hourly. In this case, they are using the term *real time* to mean "faster than before."

Here, we define *real time* as being available in time increments much less than a second. At this level, there is a significant jump in the cost of providing the data. Think of needing to handle a stream of data. You must have hardware, software, and a staff that is capable of handling the data this quickly. There are likely other jumps in cost as you go from daily data to hourly to data by the minute. But, from an IT point of view, if you have extra seconds of time, you can much better spread the load on the hardware and don't require extraordinary software solutions.

With this definition, the desire for real-time data is balanced by the real cost of providing that data. We heard a talk from an analytics professional at Netflix who ran into this trade-off. The Netflix business managers who wanted the real-time data weren't necessarily the ones who would have to pay for the IT infrastructure. So to help make the decision, Netflix analytics professionals created a litmus test: Netflix would only consider providing real-time data if the management team was willing to allow algorithms to make decisions on the data in real time.

This was a strong test. Where management was going to allow algorithms to process and act on the data, Netflix would consider it a candidate for real-time data. If management wanted to have a person look at the data and make a decision, Netflix wasn't going to invest in providing the data in real time. After all, what was the value of real-time data if there was a delay in analyzing and acting on the data?

So no matter how you define *real time*, this discussion shows that it is important to understand the cadence of the decision. If decisions are made once a day, you may only need data that is available daily. You don't want to over-invest in IT solutions that provide more than is needed.

Finally, one of the problems of monitoring real-time data is reacting too fast to variability and not problems. That is, if your real-time system gives you a data point that could signal a problem, it might turn out that it was just a bad reading or simply an anomaly. And if you were to fix the "problem," you would likely only make the system worse, since there was no real problem. The manufacturing sector figured this out a long time ago with control charts. A control chart is a tool developed to monitor quality issues and alert managers if something is wrong in a process. A control chart defines the normal measure for the machine or part. If your measure is equal to the normal measure, everything is okay. More importantly, the control chart provides upper and lower limits around the normal measure to indicate that it is also okay for the measure to vary from the normal measure. So if a particular reading isn't exactly normal but is within the upper and lower limits, you're still okay. Measurements are taken from a process over time and plotted on a chart. You take action only when you see abnormal variance. The abnormal variance depends on the situation but could involve measures that go above the upper and lower limits (or multiple measures above and below), or too many consecutive measures either below or above the normal measure. Abnormal variance indicates a possible problem that should be investigated. If you react to the normal variance, you will spend time "fixing" something that isn't broken.

A good example of this comes from *Blackett's War* and the use of analytics in Britain in World War II (refer to Chapter 3, "The Analytics Mindset"). Once the British had set up their radar system to the anti-aircraft guns, they begun to track the number of bombers shot down each night. And it was a serious concern if the Germans changed tactics to make the system less effective. This would have a very negative impact on Britain's ability to defend its cities. It would also require the British to modify their system. Anyway, according to the book, the average number of bombers shot down per night over London was about five. After one night of heavy bombing, the Brits had shot down only two. The general was questioning whether they needed to change tactics. The analyst had to explain to the general that one night with only two downed bombers was within the expected range. He did not want to change tactics for the next night and risk lowering the actual average. He said it was better to look at the pattern of the data and react when they had a significant reason to believe the tactics were no longer working.

Descriptive Analytics Through Visualization

Any good business intelligence system has significant data visualization capabilities embedded in it. You can't just explore and report on your data in table format. You need the ability to chart, graph, and plot your data. You need this so you can understand the data and so you can share your findings with others. Human beings look for patterns, and these patterns are often not evident when we simply look at data in tables. The right visualization may allow you to gain deeper understanding in a much quicker time frame. Just looking at data in tables will not lead to good descriptive analytics. In fact, many business intelligence systems sell themselves on their ability to help you visualize the data. The back-end data warehouse may make the system work, but it is the visualization capability that sells the system and where people spend the most time.

But visualization isn't just window dressing. We've given visualization its own section because it is so important to the field of descriptive analytics, and it goes far beyond just business intelligence. Interesting research suggests that it is possible to understand some large data sets only through visualization. The company Ayasdi is using machine learning, statistics, and the mathematical field of topology to help visualize large data sets in new ways. The company claims that by using its visualization approach, it was able to unlock some new insights in a large database of cancer patients that had been around (and extensively analyzed) for several decades. It was only through visualization that the secrets were revealed.[5]

Ayasdi's educational material gives a great example of the power of visualization. The primary claim is that people have a great ability to understand visual data. Therefore, Ayasdi has worked hard to present data in a visual manner. The company explains it like this: If you look a messy garage packed full of all kinds of different items, you have the ability to look into that garage and immediately spot the basketball. That is the power of human visual capability: We can quickly sort through an enormous amount of visual data and find patterns.

As another example, remember John Snow's map of the 1854 cholera outbreak in London from Chapter 1? Presenting the data in a visual manner helped convey the true story. So, in essence, the reason Snow's descriptive analytics worked was because the visual evidence was so strong. The map also allowed him to convey a lot of information in a small space.

Snow's map was powerful, but a more famous chart, shown in Figure 5.9, of Napoleon's invasion of Russia, is often used to kick off a good discussion of visualization. Even though this map is not written in English (you can easily find a translated and easier-to-read version with a quick internet search), you can tell that the chart shows the number of troops in Napoleon's army (this is the thickness of the line) as it marches east to Russia and then back to France in defeat. You can see the line starts out very thick on the far left. By, the time the army gets to Moscow, it is a fraction of its earlier size. And you can see that it shrinks even further as the army marches back west. By the time the army arrives in France, it is a mere pen stroke width. In addition, the artist has drawn the rivers the army encountered. This gives us further insight because it shows the toll each river crossing took on the army.

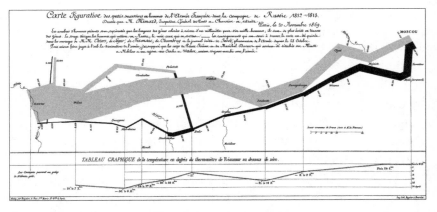

Figure 5.9 Original Map of Napoleon's 1812 Invasion of Russia (Made in 1869)

We've shown and discussed some complex and powerful visualizations, but the basics will also get you far. Visualization starts with the ability to turn your tabular data into a graph. Figure 5.10 shows how turning a simple table listing store sales into a bar graph makes it easier to see which stores are selling more and the relative difference between the stores.

Store	Sales ('000s)
A	$2,400
B	$2,185
C	$2,075
D	$1,850
E	$1,810
F	$1,560

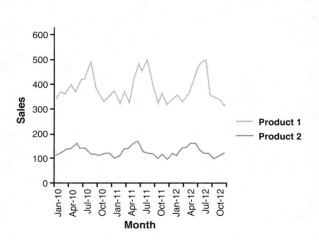

Figure 5.10 Comparing a Table of Data to a Bar Chart

As another example, Figure 5.11 shows a simple time series line graph of sales by month for the past three years for two different products. When we plot this data on the same graph (here we show only half of that data in the table[6]), we are able to detect the fact the products have slightly offset seasonal highs and lows.

Month	Product 1	Product 2
Jan-10	340	110
Feb-10	370	120
Mar-10	360	140
Apr-10	400	140
May-10	370	160
Jun-10	420	140
Jul-10	420	140
Aug-10	490	120
Sep-10	390	120
Oct-10	360	110
Nov-10	330	120
Dec-10	350	120
Jan-11	370	100
Feb-11	320	110
Mar-11	370	140
Apr-11	330	140
May-11	420	160
Jun-11	480	170

Figure 5.11 Sample Line Chart Showing Sales for Two Products (Product 2 is the lower line)

In addition to using the basic charts and graphs (and there are a lot more of them than we highlight here), people are using more maps to convey information. This can be a powerful way to visualize information. For example, Figure 5.12 shows a map of the same data shown in the bar chart in Figure 5.10, you can start to see more patterns immediately. You can now see that the two stores in downtown Chicago (A and B) are two of the largest. With the exception of store C, the stores in the suburbs are selling less than the Chicago stores. Mapping the data in this way allows you to convey a lot more information.

Figure 5.12 Store Sales Plotted on a Map

Finally, a trend in visualization that is catching on is the use of infographics.[7] You've likely seen infographics alongside articles in magazines and newspapers. An infographic is a single picture that uses various charts, graphs, and pictures to convey information about a subject.

From a management point of view, the good news is that technology has made creating graphs, charts, and maps much easier. And if your visualizations are online, it is easier to make them dynamic and interactive. This means you should expect better descriptive analytics with visualization. You should be able to create these visualizations yourself, and your teams should create them as well. Using visualizations should become a key part of your efforts to do good descriptive analytics.

If there is a downside to technology allowing us to more easily create better visualizations, it is that the visualizations become art rather than convey information or insight. Although beautiful visualizations are pleasing to the eye and draw immediate praise, they fail if they don't help with descriptive analytics. To prevent this problem, managers need to understand what visualizations are for and the fundamental principles of visualization.

Of course, whenever you talk about building a good visualization, you are likely to have reasonable people disagree. Different people prefer different types of charts, different colors, and different layouts. We won't get into this level of detail. Instead, here we'll focus on the main principles and ideas behind good visualization.

Let's start by thinking about the purpose of visualizations. Michael Schrage, in his *Harvard Business Review* blog post, argues that visualizations should be created to provoke dialog, discussion, and interaction.[8] He goes on to argue that you shouldn't just ask how to best present the data. Instead, he says you should determine what kind of conversations you want to start or how you want people to interact with the visualization. He is aiming for creating visualizations that create new value by creating the right interactions between people. He also pushes for interactive visualizations that allow people to drill into the data as their conversations evolve. Another way to think about this is that descriptive analytics through visualizations should make people aware of the issue and provoke interesting discussions about what to do about it. Ultimately, this should lead to better decisions. He wants these graphics to help us gain insights we wouldn't otherwise have and that can't be presented as static charts.

Justin Holman, who holds a Ph.D. in geography and is CEO of TerraSeer, a mapping company, complements Schrage's view with his view of maps as portals for business discussions. Holman argues that too many businesses present data in Excel tables or in simple charts that could be plotted on a map. He argues that the tables and charts sometimes stymie conversation by making it difficult to get through the jargon or understand the comparisons. In contrast, the same data presented on a map allows greater participation from more people. The map conveys more information and, more importantly, presents information in a way that is easier for more people in the organization to grasp. This invites more people to join in the discussion. For example, the people in the field for a company may be more engaged when they see data presented at the geographic level. Since they are in the field, they can more directly comment on what they are seeing. Holman argues that it is no mistake that military campaign planners use maps (and not spreadsheets) to plan battles. A map conveys more information and allows more people to comment and make suggestions—helping prevent costly oversights and missed opportunities.

In terms of specific principles of good design principles, there are many great sources. Probably the most famous is Edward Tufte. His 1983 book *The Visual Display of Quantitative Information* is a classic in the field. From his work, we get several guiding principles for good visualizations:

- **A visualization should convey many ideas in a short time and with as little ink as possible in the smallest possible space**—Bill Franks amplifies this comment in his *Harvard Business Review* blog post when he says that that a good visualization conveys information immediately.[9] That is, when someone looks at a visualization, she should immediately understand it. He gives an example of a social network graph that shows a single person connecting many different subgroups. We can immediately understand that this person would be key to reaching those subgroups.

- **A visualization should tell the truth about the data**—Managers often encounter bar charts that have a y-axis that doesn't start at zero. This distorts the relative difference between the values being compared. Figure 5.13 shows the same store data presented earlier in this section, but here the y-axis does not start at zero. This makes the sales at store A seem more than twice as large as those at store D and more than 10 times as large as those at store F.

Store	Sales ('000s)
A	$2,400
B	$2,185
C	$2,075
D	$1,850
E	$1,810
F	$1,560

Figure 5.13 Store Sales Data from the Figure 5.10, Here with a Large "Lie Factor"

Tufte calls this the *lie factor*. He defines the lie factor as the ratio between the effect in the visualization and the effect in the underlying data. That is, you physically measure the difference in the ink used or size of the graphic and compare it to the actual data. So if the lie factor is close to 1, you have accurately represented the data. If it is far from 1, your visualization is distorting the true data. You typically run into trouble with the lie factor if you use bar charts where the axis doesn't start at zero, or if the area of the pictures you draw to represent data doesn't accurately reflect the underlying data points.

The lie factor doesn't have to be intentional. The tilted 3D pie chart is a good example of this. It is easy to create a nice-looking tilted 3D pie chart using Excel. However, this type of chart increases the lie factor. When a pie chart is tilted to add the 3D effect, some parts of the chart become relatively small. If you want to mislead with a pie chart to make an unfavorable slice look larger, you can put that slice in the front of 3D pie chart, and it will look much larger than it should.

The pie chart suffers from another visualization problem: It is hard for people to compare areas. (In fact, with a quick internet search, you can find many people who would just prefer to do away with the pie chart because it is misleading.) Another problem with pie charts and area charts is that people tend to compare heights and not area.[10] So, based on what people are looking at, they may be creating a lie factor even if you were careful to avoid it.

You usually run into the lie factor when you start to make complicated visualizations and infographics. And that's also where you run into the problem of turning an infographic into art rather than an example of good descriptive analytics. So a lot can go wrong if you don't have any guiding principles.

For some good management principles on infographics, we turn to Alberto Cairo, author of the book *The Functional Art*. He claims that a good infographic should present the data (and maybe several different variables), allow easy comparisons, help organize the data, and help make correlations apparent. We find it very interesting that the top item in the list is something as seemingly obvious as presenting the data. If an expert has to put an obvious item like this in a list, it indicates that a lot of infographics end up hiding the data in the graphics or create a lot of visual noise to show just one variable. Cairo's other three points are somewhat related. If you have interesting drawings and graphics in your infographics, it is tempting from a layout point of view to put data or charts where they fit. But, you should be careful that you don't let the layout requirements of the drawings separate your data. You should make sure someone can easily look at the data and make comparisons of like numbers and see correlations between related variables.

Descriptive Analytics Through Descriptive Statistics

When analyzing data and doing descriptive analytics, you shouldn't forget the basic lessons from statistics. If you have a set of numeric data, it can often be helpful to calculate basic facts about the data.

Two of the most basic but important statistics ideas are the mean and the standard deviation. The mean, or the average, gives you a sense of the midpoint of the data. The standard deviation shows the variation in the data. A low standard deviation (relative to the mean) tells you that most of your data points are close to the mean. A high standard deviation tells you that your observations are widely scattered around the mean. There is a convenient way to compare the variation of two data sets: the coefficient of variation. To calculate this, you simply divide the standard deviation by the mean. If you get a number close to 0, it means you have little variability. A number close to 1 typically indicates quite a lot of variability, but the number can go even higher. It is also often very useful to understand the minimum, maximum, and quantiles (for example, quartiles, quintiles, deciles, and so on).

For example, think of a retailer that is comparing the sales figures for its stores. In the state of Michigan, the mean sales for the 50 stores are $1 million per store, with a standard deviation of $175,000. In the state of Florida, the mean sales are $1.5 million per store, with a standard deviation of $650,000. So what does this tell you besides the fact that the stores in Florida sell more? It tells you there is a lot more spread in sales in Florida. The coefficient of variation for the stores in Michigan is 0.18, while it is 0.45 in Florida.

If you want to visually explore the variation, you can use a histogram, which is a great, underused report. Figure 5.14 shows the number of stores in various sales brackets. You can see that most stores in Michigan are close to the mean, while Florida has a lot of big and small stores. From a management point of view, a high coefficient of variation tells you that there might be more to the story and more to investigate.

In addition to using the mean, standard deviation, and coefficient of variation to compare different columns of data, a powerful and somewhat overlooked calculation is the correlation matrix. This is a standard calculation in statistics and is even part of the Analysis Toolpak that comes with Excel. The correlation matrix allows you to quickly see how different columns of numeric data are correlated. For example, say that you have a data set of about 250 men that holds data on their body fat, age, weight, height, and the measurements of their neck and abdomen. Exploring the correlation between any two different fields may give you insight. A correlation matrix, as you can see in Figure 5.15, is easy to read. If the number is close to 1, it means the data in these two columns moves together: When one goes up by a certain percentage, the other one does as well. When the number is close to 0, it means the data in the two columns moves randomly and not in any kind of correlation. When the number is close to –1, it means the data moves in exactly the opposite direction: When one field goes up, the other goes down. For example, we see that body fat and the ab measures are 81% correlated, which means these two values are highly correlated. The higher the ab measures, the more body fat someone has. You can also see that the measurement of the neck is only 49% correlated with body fat. And the height is hardly correlated at all; if anything, taller men have a tiny tendency to have less body fat.

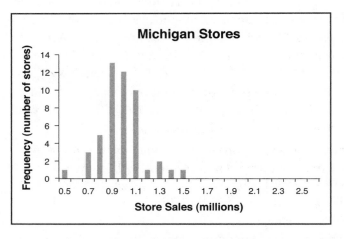

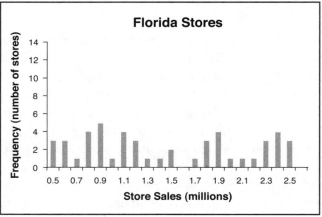

Figure 5.14 Histograms for Store Sales in Michigan and Florida

	Body Fat	Age	Weight	Height	Neck	Ab
Body Fat	1					
Age	29%	1				
Weight	61%	-1%	1			
Height	-9%	-17%	31%	1		
Neck	49%	11%	83%	25%	1	
Ab	81%	23%	89%	9%	75%	1

Figure 5.15 Results of Correlation Analysis for Body Fat

In the book *Big Data* that we mentioned in Chapter 2, "What Is Driving the Analytics Movement?" Mayer-Schonberger and Cukier claimed that with Big Data, you no longer needed to know why something caused something (causality); you only need to know that the data was correlated. This is one quick way to do the correlation they we just talked about. And it is a good way to better describe the data you have.

Descriptive Analytics Through Machine Learning

Up to this point, we've described descriptive analytics as a very deliberate process. You have a set of data (whether you created a data warehouse or just have access to data), and you use that data to run SQL queries or build visualizations. But you are asking the questions and seeing the results. In some cases, you are asking the same questions every time (like when you create standard reports). In other cases, you are doing ad hoc analysis.

However, in some cases, you don't even know enough about the data to do ad hoc analysis. That is, you don't know the questions you want to ask or the answer to the question is so complex that you couldn't write a SQL query to get the answer. This is where machine learning can come in. And now you can see why machine learning algorithms grew out of the field of data mining. Just like gold miners, we are searching around in the data to see if we find something of interest.

The most obvious cases of descriptive analytics through machine learning come from unsupervised machine learning. That is, we have a set of data, and we want to better understand it. Earlier in this chapter, we talked about the use of topology to visualize a large data set; this is unsupervised learning. In Chapter 4, "Machine Learning," we talked about the clustering and association rules (such as market-basket analysis). These algorithms help you describe a data set in new ways. You may uncover information you did not already know. Or you might find patterns that were obvious to everyone before the exercise (perhaps the fact that people who buy cereal also buy milk). Or you may not find anything. This is both the beauty and the challenge of unsupervised learning.

In addition, some supervised learning algorithms can help you better understand your data. Think back to the decision tree example in Chapter 4. That example used a decision tree to help predict whether someone in their 20s would accept a marketing offer. As a side benefit, when the algorithm built the tree, it also helped us better describe the data. It broke up the data set into a tree that we could then analyze outside just our narrow interests in the marketing offer.

So even though we have nice categories for descriptive, predictive, and prescriptive analytics, keep in mind that this is only a framework for helping think more clearly about the field. In reality, the lines between the different types of analytics can blur. And, certainly, some algorithms can span different types of analytics.

6

Predictive Analytics

Predictive analytics is broadly defined as techniques that help us make predictions about the future or about new data observations.

For example, predictive analytics has long used information based on past sales and promotions to predict sales next week, next month, or next year. Extensions of this model might also include data about competitors, general market trends, macroeconomic activity, or the weather to help predict future sales. Also, predictive analytics can help you test a hypothesis with a sample of data. For example, will sales increase by 10% if you lower prices by 5%? Or if your website has pictures of people, will people stay on your site for an extra 10 minutes? Machine learning algorithms are pushing into new areas for predictions. For example, predictive analytics can help determine if a new email is spam, if credit card activity is likely to be fraud, or how well a patient may respond to a treatment. Also, predictive analytics, using simulation, can help determine how a system may evolve over time.

As mentioned earlier in the book, predictions can never be guaranteed to be accurate. But good predictive analytics will help you quantify how confident you can be in an answer. The concept of confidence is a powerful management tool. When making decisions, you want to understand the importance of your decision and the confidence in your predictions. For example, if you have an important decision but the results from your predictive models have low confidence, you may want to proceed with caution or gather more information.

In this chapter, we take a look at some of the most common types of predictive analytics. We'll start with the area that has been around the longest: forecasting and regression. Then we'll revisit the machine learning algorithms we covered in Chapter 4, "Machine Learning."

We'll also spend time discussing a popular method for improving websites: A\B Testing. Finally, we'll talk about an important niche area of predictive analytics: simulation.

Forecasting with Regression

One of the most common and longstanding uses of regression analysis in companies has been to help forecast demand for their products or services. *Regression* simply refers to the process of trying to predict the value of one variable (the dependent variable) based on the values of other variables (the independent variables).

The minimum requirement to do such an analysis would be to capture and load historical sales data into a regression model (in machine learning terms, this is the training data). Historical sales data is time stamped (usually by week or by month) and may be augmented with information about special promotions or events. Typically, the models will perform better if you have several years of data. That is, you would like enough data to see the same month or annual promotion play out over several years. The regression model then returns a simple equation that allows you to plug in the future month and future promotions and come up with a prediction for the sales in that time period. The equation automatically accounts for the overall trend in the data, seasons, and promotions. The equation also provides you good information on how confident you should be in the predictions. This is often called the *forecast error*, and it is commonly presented as upper and lower bounds. For example, the result of the forecast may predict sales of 100, with a 95% confidence that the true value will fall between 80 and 120. If the 95% confidence interval comes back with 10 to 190, you would be wise to realize that you shouldn't be confident in the forecast.

People may not realize that regression is used for this type of predictive analytics because it has been embedded in demand planning systems. Most firms have demand planning systems that allow them to track historical sales, promotions, and events and then return the forecast results—which come from regression. These systems can also allow you to share information in an organization and collaborate with others. For example, besides relying just on historical information, you might want to have your sales team, your suppliers, or your customers provide information into the system to make it more accurate.

Even though forecasting is a well-developed field, the rise in data has had a big impact.

In the past, firms relied on weekly or monthly data. New forecasting systems are being built to take daily data from stores to better predict demand at the store for the next several days. The idea is that this information can help you get the product you need to the right store.

Also, companies are realizing that they can better forecast when they incorporate data sources from outside the firm. That is, in the past, companies just used their own historical data to predict the future. But as more data became available from external sources, companies realized they could leverage that data to make better forecasts. For example, an auto-parts store could use the vehicle registrations in its market to make its predictions better. Knowing the types and ages of vehicles near the store could help predict demand. The same would apply to the demographics of a store or adding information about traffic flows or competitor locations. For short-term forecasts, many retailers have found a lot of value in considering weather forecasts to predict in-store sales.

Machine Learning and Ensemble Models

Although forecasting is very widely used, it has been around too long to get a lot of press. Instead, newer fields like machine learning—a new area, using newer algorithms, and applying it to new data—get the majority of the press attention.

We covered machine learning algorithms in Chapter 4. We could have also included the information from that chapter here because the supervised machine learning algorithms form the heart of some of the new ideas in predictive analytics. However, since machine learning algorithms are also used in descriptive analytics (for example, with unsupervised algorithms to uncover insights or using a decision tree to describe the data), we decided to cover it earlier.

As a reminder of how supervised machine learning algorithms are part of predictive analytics, think about how the decision tree and k-nearest neighbor algorithms give us ways to predict an outcome for a new observation. Also, although we limited our discussion of regression in the previous section to forecasting, regression can also be used more generally to make predictions about data. For example, if you have data on the selling prices for homes in a particular geography, along with the square feet and number of bathrooms, you can predict the future selling price for a house that is 2,500 square feet and has three bathrooms. This is not forecasting but a use of regression in line with other machine learning algorithms. It is essentially what services like Zillow do.

Remember that there are probably hundreds of these supervised machine learning algorithms, and each one has a variety of parameters that you can adjust. It is beyond the scope of this book to cover all the combinations. But as a manager, you should know that these choices exist.

So why do all these choices matter? What people are finding in practice is that it is difficult to determine a single best algorithm for forecasting. This should not be surprising: Making predictions is hard (especially about the future, as Yogi Berra said). But people are finding that instead of spending a lot of time finding the best algorithm, they can be more effective by running their data through a lot of different algorithms with different settings to see which ones are the best. A related strategy is to use some combination of the algorithms to make predictions. Bill Franks, in his book *Taming the Big Data Tidal Wave*, describes these as ensemble models and compares them to crowdsourcing for models—that is, you rely on the wisdom of a lot of models rather than just one. And even if you have found the best algorithm now, it doesn't mean you can't make adjustments over time. In fact, the best of these systems recognize that the world changes over time, and the algorithms adjust over time.

Complicated machine learning systems like IBM's Watson or Aysida's system for visualizing data that we discussed in Chapter 5, "Descriptive Analytics," do not rely on one algorithm. When you read about these systems, you realize that they rely on hundreds (or thousands) of algorithms. Then they have other algorithms to decide which of the answers are best. These two extreme examples of ensemble models give you a good idea of how they work.

Most of us won't be managing systems as complicated as the ones mentioned above (although we will be using them more frequently). But the same ideas apply to other areas. Often, we see people spending a lot of time trying to pick the single best algorithm. This approach probably comes from our training in schools and universities that there is a single right answer and from the fact that computers used to be much slower and therefore didn't give you the ability to try different things. Instead, the new trend in analytics and the use of machine learning is to try a lot of different algorithms and see what works best. And as you get better at this, you can systematically try a lot of different algorithms and let other algorithms pick the best answer. We mentioned the use of regression for forecasting demand at the start of the chapter. Many different regression models and assumptions are used. And, in practice, forecasting systems have long tried out many different regression models on the data and selected the best. They selected the best using statistical techniques.

The lesson for predictive analytics is that a single machine learning algorithm can help you make predictions you couldn't before, but using many different machine learning algorithms can help you make even better predictions.

A/B Testing

In Chapter 2, "What Is Driving the Analytics Movement?" we mentioned that the idea of testing business ideas (rather than just talking about them) helped ignite the analytics movement. A/B testing is at the heart of this testing. Although A/B testing can apply to many areas, the field of analytics almost exclusively uses the term to refer to testing the performance of websites.

So what is A/B testing? The idea is very simple. You take your existing website as version A. You then make a small tweak to this site and call the new version B. The tweak should be small because you want to isolate the impact of this one change. For example, you might change the color of a button from blue to orange, you might change the text on the button from Purchase to Buy Now, or you might put a new picture on the site. At the same time, you decide what you want to test for. Usually you test to see if more people purchase a product, stay on the site longer, or sign up for more services.

To run the test, you simply set up your website so that you show a random number of visitors version B and the others version A. You then collect data on what the visitors buy, how long they stay, or whether they sign up. At the end of the test, you run statistical analysis to see if version B led to better outcomes. If it did, version B becomes your new default website.

To run a good A/B test, you need to make sure you have enough visitors, you randomly assign versions A and B, and you run the test for a long enough period. You need enough visitors to make sure you have a statistically significant sample size. If your sample is too small, the different outcomes may just be due to chance, and you can't really say that you've found a true difference between version A and version B. It is also important that you randomly assign visitors to version A or B. You don't have to *equally* assign A and B. In fact, you may want to show only a small number of visitors the new site. But by *randomly* assigning visitors, you remove biases that may creep into the analysis. For example, you don't want to just show visitors from a Florida IP address the version B; they might be fundamentally different from the rest of your visitors. And, related to random samples, you want to test for a long enough so that you don't get biases based on the time of day or the day of the week—your 8 a.m. visitors may be different from your 9 p.m. visitors or your Monday visitors different from your Saturday visitors. When you research A/B testing, you are likely to find stories of firms seeing such conclusive results within a few hours that they make a switch. You have to wonder if such a story was just a dramatic embellishment or if the firm risked making a change based on a time-of-day bias.

You can see why large, high-volume websites like A/B tests: They can quickly run a lot of tests. When you read about a company doing 10,000 tests per year, you can assume that they are running tests like this.

Many web developers, software engineers, and internet entrepreneurs suggest that A/B testing is something that every website should do—at least every website that cares about doing better. And the evidence suggests that many websites in fact take this advice.

Although early web companies have been using A/B testing for a long time, it seems that the 2008 Obama campaign brought the technique more popularity.[1] (We discussed this example in Chapter 1, "What Is Managerial Analytics?") Obama's campaign used A/B testing to select the right image for its website and the right text on its button to increase the signup rate by 40%. This signup rate increase translated into millions more volunteers and millions more dollars in fundraising. Since this was disclosed and people could relate to the results, people started to think about using the same techniques for their businesses.

After the election, the person who ran Obama's A/B testing went on to start a company called Optimizely that allows firms to more easily run A/B tests. Before that, someone had to code up both websites and the tests. Optimizely took the programming out and allows for faster tests.

A nice *Wired* magazine article that discussed the Obama campaign's use of the A/B testing also brings up some other interesting points about A/B testing in general[2]:

- **A/B testing requires a change in how decisions are made**—No longer should the experts select the colors, text, or pictures for a website. The experts suggest a reasonable set of alternatives, run the tests, and let the results of the test determine what to do. This can mean a big change in how decisions are made. You can imagine an awkward conversation where President Obama recommends a certain picture for his website, and the A/B tester has to break the news that a different picture should be used!

- **As a result of the tests making the decisions, there is a feeling that people no longer care why one choice was better than another**—The *Wired* article discusses the fact that people are reluctant to draw conclusions from a test because next month an opposite color, word, or picture may lead to an increase in sales. But we've seen other cases where enough tests come back with similar results to draw conclusions. For example, maybe large pictures help drive traffic. In any case, A/B testing changes how you draw conclusions about your customers. Maybe your customers change frequently, and you use A/B tests to keep up with the changes or maybe they don't, and you use prior tests to make future designs better.

- **A/B testing might lead you down a path to improve your existing site but miss the benefits of a completely new site**—Part of this problem comes from the fact that A/B testing is often called "website optimization" in technical and computer engineering circles. This is a misleading name. It implies that A/B testing will give you the best possible website. It doesn't do this. And this is a misuse of the word *optimization*, which we better explain in Chapter 8, "Prescriptive Analytics (aka Optimization)." What A/B testing *does* do is to predict which of two different designs will work better. And it does a good job when you make incremental changes because it helps isolate the changes. Remember that it is not trivial to do major changes to your website. For example, if a visitor sees version B with a green button and then comes back later and sees the same button colored red, it will not be a big deal. However, if version B is completely different and it gets rejected, the same users may be more confused when they come back and see the old design again; this can have serious implications for the brand. So testing dramatically different designs is probably not appropriate for A/B testing. The trap with incremental changes is that your initial design may not be very good. This is not a problem with A/B testing, but it's something to keep in mind if someone mentions that A/B testing will "optimize" your website. It can make it better, for sure. But you still may face tough business decisions about when a more significant redesign of the site is warranted.

One company we know of warns that A/B testing applied to all visitors is only good for coming up with a design that is good, on average, but may not be great for anyone. A better approach might be to design different landing pages for different visitors, depending on their characteristics. These different landing pages can themselves be A/B tested for improvements. You can see that there are a lot of management decisions you still need to make.

Isn't This Just Hypothesis Testing?

If you are new to A/B testing but have some background in statistics, you may be thinking that A/B testing looks a lot like hypothesis testing. Strangely, in all that's been written about A/B testing, this connection is almost never brought up. We've concluded that A/B testing primarily is used by computer engineers (who know a bare minimum about statistics) who don't necessarily draw the connection. As you've seen in this book, this is a common issue that makes understanding the field of analytics so confusing. A group latches on to a technique, it becomes a powerful and beneficial technique, the group labels what they are doing as "analytics," and there is no reason to point out that the technique is part of a larger body of knowledge.

A/B testing is a type of hypothesis testing. It is good to understand this so you can apply the techniques to more decisions and so you can apply lessons learned from hypothesis testing to A/B testing.

In traditional hypothesis testing, an alternative hypothesis is tested against a null hypothesis. Essentially, a *null hypothesis* represents your default position—that is, you will believe the null hypothesis to be true unless there is sufficient evidence to change your mind. The specific idea that you are trying to test against the null hypothesis is known as the *alternative hypothesis*. In A/B terms, website B is part of the alternative hypothesis—for example, say that you want to show that changing the words on the button to Buy Now will lead to more sales. The null hypothesis is simply that version B (where the wording has changed) will be no better than version A in generating sales. Or, as another example with which people are very familiar, the U.S. criminal justice system creates an alternative hypothesis that the defendant is guilty. The null hypothesis is that the defendant is not guilty.

Then, in traditional hypothesis testing, you collect data and analyze the data to see if you have enough evidence to reject the null hypothesis. By rejecting the null hypothesis, you accept the alternative hypothesis. In this chapter's example, you analyze the data and see if you can reject the claim that version A will lead to at least as many sales and version B, and, equivalently, say that version B will lead to more sales than version A. In the criminal justice example, you analyze the data to see if you can reject the claim that the defendant is not guilty and therefore conclude that the evidence leads you to say that the defendant is guilty. One thing to note is that our choice in this type of analysis is to either accept or reject the alternative hypothesis. Accepting means believing the alternative to be true, but rejecting does not mean believing that the alternative is necessarily false; it means you don't have sufficient evidence to accept it as true. That is, a not guilty verdict does not mean innocent.

A lot of management insight that is built into accepting or rejecting the null hypothesis gets lost in some of the A/B discussion.

First, there are two types of errors you can make when accepting or rejecting a hypothesis. By tradition, a type I error is finding the alternative hypothesis true when it is really false. For example, a type I error would be your data showing that version B is better when it really isn't or a trial finding the defendant guilty when he is really is innocent. A type II error is the opposite: You don't reject the null hypothesis but should have. For example, a type II error would be your data showing that site A is just as good as B when it is really inferior or the trial finding the defendant not guilty when in fact he is guilty. So, as a manager, you should remember that you are balancing these two types of errors.

Second, although there is no getting around type I and type II errors, you should adjust your acceptance criteria based on the implications of the types of errors. This is easily seen in the criminal justice example. It is accepted that it is much worse to convict someone who is innocent—which would be a type I error. Therefore, the standard in a criminal case is "beyond a reasonable doubt." In other words, you want to really work hard to avoid type I errors. For quantifiable statistics tests, you generally set a threshold of 5%. This means you are willing to reject the null hypothesis and accept the alternative hypothesis if there is a 5% or less chance you are making a type I error. But if you could quantify "beyond a reasonable doubt," you might say it is 1% or 0.1%. Now, when making less serious decisions, such as changing the words on a button on your website, you might want to stick with 5% or you might go as high as 10% or 15%. The management point here is that you are not simply testing whether B is better than A; you are seeing if it is statistically better. And, since there may be an assumption that changing the website introduces some hidden costs, you might want to set a relatively tough (5% or 10% and not 50%) threshold for accepting that version B is the right way to go.

To minimize type II errors, the best solution is to collect more data. And, to echo an earlier point, you want to make sure the data is unbiased. So when doing hypothesis testing, it is always possible to defer a decision and collect more data to see if an answer becomes apparent.

To wrap up, A/B testing should be part of your analytics toolkit. But don't forget that A/B testing is just one type of hypothesis testing. The more decisions you can make with good hypothesis testing, the better off you will be.

Simulation

Like forecasting with regression, simulation is often forgotten as a powerful part of predictive analytics. Simulation technology has been around for a long time, is commonly taught in universities, and is supported by a wide range of software vendors. The term *simulation* in one sense is a rather generic term that is often used to describe any situation in which people use computers (or other means) to try to emulate, test, or understand a real-world system or process. This can often lead to confusion between terms like *simulation* and *optimization*. We prefer to describe this process as *modeling*, with *simulation* and *optimization* being different types of analytic models.

Simulation models are defined by the fact that they involve sampling numbers from probability distributions that will then drive the behavior of the system or the value of the output being tracked. Since the model draws on random variables, any one run of the model is a result of the random variables that happen to be selected. So, to better understand the system, you run simulations multiple times so that you get a wide range of random variables. Then you look at the performance of the system over the multiple runs to get a sense of how the system will behave over time. The primary motivation for using simulation models is that there are many real-world problems where there is a high level of uncertainty or randomness in some critical elements, and there are no known mathematical techniques (or closed-form equations) to precisely predict the performance of the system or the solution value. In these cases, you essentially use simulation models to conduct experiments. By conducting these experiments repeatedly and in a clever way, you can develop confidence in your conclusions.

For example, if you work for Disney, you may want to create a simulation to see how long the lines will be for a certain ride. The simulation model of this ride would include a lot of parameters that would be random. The number of people arriving to the ride would be variable (sometimes people arrive in big groups) and would change throughout the day (fewer people may arrive at lunch time). The amount of time that it takes people to load and unload from the ride is also variable—sometimes it takes longer or shorter amount of time. And, since friends and family members ride together, you might not be able to fill up all the seats available. Maybe the ride itself has some randomness in the time to complete. The simulation model is run multiple times (actually, the software takes care of this for you and is run maybe 10,000 times) and accounts for all these factors and returns a result showing the average number of people in line and their wait times throughout the day.

At this point, you may be thinking that you could just average out the random variables and calculate the average time directly without the bother of using a simulation model. But it is the variability that makes the problem challenging. It is times when a lot of people arrive to the ride at the same time, when a lot of seats go unfilled, and the load times happen to be long that the lines will really build. And Disney wants to know about these problems. If you averaged everything out, Disney would lose this insight.

Also, you might be wondering why simulation is classified as predictive analytics. You could make an argument that it is more descriptive—that is, it describes how a system behaves. In some ways this is true, but the true power of simulation focuses on predicting how a system will behave in the future, given different parameters. Understanding the current system may be very useful to Disney in identifying where there are problems the company would like to address (for example, which rides seem to have problems with lines becoming too long). But what Disney may very well like to do next is to test different ideas that might improve the situation (adding an additional employee, changing the loading or unloading process, and so on), and using a simulation model to predict the system behavior under various conditions is where the real value lies.

Further, you may wonder why we don't classify simulation as prescriptive analytics. Doesn't the simulation lead to telling you what actions to take? Again, there is a fine line. In some cases, the results of the simulation may make the best solution obvious. But a simulation run does not actually tell you what you should do. It just predicts the outcomes. Let's look again at the Disney example: Disney probably wants to do something if it finds long lines. It can use its simulation model to see what would happen if it changed the loading process to fill up more seats or make the process faster. Disney then runs the simulation model to see what would happen under a number of scenarios, and it picks the one it likes. In this case, the model is not telling Disney what to do; rather, Disney is giving the model various choices, and the model predicts the outcome for each of the different choices. This can be confusing because many simulation packages come with an "optimization" feature that conveniently allows to you to select various choices and have the model automatically run each of them and tell you which one is best. This is essentially optimization through enumeration rather than application of specific optimization algorithms. This approach can work very well when the number of available decisions is relatively small and the range of choices within that decision is relatively narrow. However, this approach will fall down when confronted with a problem that has a very large number of possible choices. Suppose, for example, that you were confronted with the task of scheduling 25 tasks, each with an expected time to complete and some variability in that time. Your job is to determine the sequence of the tasks to minimize how late any task will be (according to a desired schedule). One possible approach would be to run a simulation with every possible sequence and then collect the performance statistics to see which sequence performed best. Unfortunately, the number of possible sequences to test would be 25!, which is equal to

15,511,210,043,331,000,000,000,000

This is not a realistic number of simulation runs you can perform (even ignoring that you would want to do hundreds or thousands of runs for each). In this case, a more realistic approach might be to apply optimization algorithms (likely with fixed assumptions about the time to perform a task), which we'll discuss in Chapter 8. The optimization algorithms (perhaps employing some sensitivity analysis) would allow you to narrow down to a reasonable number of possible sequences that would be expected to perform well. You could then follow this up by performing simulation on this small subset of solutions to see how they perform, given the variability. To really determine whether a simulation or an optimization model is the best approach for a given problem you are trying to solve, you really need to ask yourself where the "action" is. If most of the complexity of the problem lies in the variable and dynamic nature of the system, while the number and range of actual choices to make are small, then simulation is likely a very good approach. If the variability of the inputs is small or negligible, but the number and range of possible choices is very large, then you are likely to be best served with an optimization approach. Those poor souls who confront both (highly variable and dynamic inputs and a large number and range of choices) will likely need to make some compromises in the approach and find a creative way of taking advantage of both techniques.

Simulation models have a wide range of applications. You may have built simulations in Excel without knowing it. If you've built a spreadsheet with the RAND() function and then pressed F9 multiple times to see what happens, you've built a simulation model. A common Excel simulation would look at likely sales or performance of a company or product. Simulation has been used in hospital emergency rooms to determine wait times for patients and to see how busy the doctors and nurses are. Simulation has been used in retail stores to see how people move about the store and how long it takes to check them out. Simulation has been used inside warehouses to understand how congested the dock doors may be. It has also been used extensively in physical sciences to gain better understanding of chemical reactions, weather patterns, drug efficiency, and so on. The list goes on and on.

One characteristic of the preceding simulation examples is that it is important to model the variability directly. When you model a hospital emergency room, you realize that it is full of variability— the arrival of patients is variable and the time to treat each patient is variable. These factors make managing an ER difficult. Simulation can help you understand the impact of variability.

To understand simulation, you should understand the difference between static and dynamic models. A static simulation is one where it is not necessary to worry about the time dimension of a system and model the interaction of specific events. For example, a simulation in Excel that determines next year's revenue using randomly generated projections for economic growth and sales team performance is static. A commonly used term to describe simulation of this type is *Monte Carlo simulation*. The idea behind this type of simulation is that there is a certain outcome that you would like to predict, such as the performance of a portfolio of investments. There might be a number of dynamic considerations in this prediction, including things like the performance of the overall market, the performance of specific industries, the performance of specific companies, currency exchange rates, and so on. Given that each of these factors is subject to uncertainty that you can, at best, try to characterize with a probability distribution, it is likely impossible using purely mathematical equations to predict the portfolio performance. However, if you knew the value of each of the underlying considerations, calculating the portfolio performance would be a breeze. So a Monte Carlo simulation will perform thousands of experiments, in each one drawing a value (based on the probability distributions) of the underlying factor, and calculate the portfolio performance of each experiment. After running enough experiments, it should provide a fairly accurate estimation of the portfolio performance (including the variability).

On the other hand, in a dynamic simulation, it is critical to consider the system over time and how various elements of the system interact. The most commonly used dynamic simulation approach is known as *discrete event simulation*. Think in this case about the model of the ER. Unlike in the Monte Carlo example, the critical input sources of variability (arrival of patients, length of care required, and so on) cannot simply be drawn from a distribution and plugged into an equation. It is critical to track those arrivals and the care provided over time. If a particular patient needs an unusually long amount of a doctor's time, then that doctor is not available for another patient. So in discrete event simulation, you are keeping track of all the resources and understanding when they are busy and when they will be free to work on something else. If this sounds like it can quickly become very complex, that's because it does. Depending on the system you are trying to model, there may be thousands of interactions to track and thousands of points of randomness; system behavior may vary widely from simulation run to simulation run. There are many powerful software solutions that will attempt to take care of the complexity for you. But you should understand that it will still take a skilled analytics professional who understands the business and modeling to feed the right model to the software and also that these models may take a long time to run—because they have to track so much data.

7

Case Study: *Moneyball* and Optimization

A really fast way to mainstream a topic that has traditionally been limited to a group of specialists is through popular culture. We see this in the culinary world—through the advent of celebrity chefs and hit reality shows. Ten years ago, if you met someone who called himself a "foodie," you would probably have looked at him with puzzlement. Similarly, a variety of different reality television shows in areas like real estate, home improvement, and auctioneering have given people a more detailed view and understanding of areas where they likely would have had little prior knowledge. For the field of analytics, perhaps the most successful example of a pop culture introduction has been through Michael Lewis's best-selling book and subsequent movie *Moneyball*, starring Brad Pitt.

The book follows the Oakland A's professional baseball team over the course of a couple seasons, focusing around their general manager of many years, Billy Beane. The A's had made the playoffs in both 2000 and 2001, but they were losing some of their top players after each season because of the inability to pay salaries competitive with what teams with deeper pockets were offering. Nevertheless, the A's won their division three of the five years from 2002 to 2006 (and finished second the other two seasons), despite the fact that their payroll was often less than half that of the competition. The argument made in the book is that this was not luck or attributed to the inherent baseball wisdom of Billy Beane or his staff but rather can be largely attributed to the organization's dedication to utilizing analytics to help guide decisions and use its limited resources in the best way possible.

Gone were the days of judging a player based on his "pretty swing" or his "great body." The A's started insisting on tangible data about player performance. Of all the major sports, baseball has been the one most heavy with statistical information. You can look back more than 100 years and find information on a pitcher's wins, losses, earned run average, and strikeouts. You would also expect to find statistics for a batter's batting average, home runs, runs batted in, and stolen bases. These traditional statistics had been used for years to judge a player's value and largely would determine how much a team would pay a player. In more recent years, a small group of statistically minded folks had made a point of tracking a much larger set of statistics. For example, a pitcher could be tracked by WHIP (walks + hits per innings pitched), while for a batter, OPS (on-base + slugging percentage) was judged a meaningful statistic. Volumes have been written about the variety of baseball statistics; in fact, the field of study is known as *Sabermetrics*, and we can consider it a very small subfield of analytics.

What Billy Beane and the A's did was to really dive into the variety of statistics available (descriptive analytics) and try to determine which statistics were most meaningful in contributing to winning games (predictive analytics). Second, they compared which statistics contributed to winning with how those statistics were being compensated in the market for players (both in terms of salary and trade value). To dramatically oversimplify, one major finding was that players with high batting averages and lots of stolen bases were likely to be overvalued, while players with high on-base averages who could hit home runs (despite possibly having low batting averages) were undervalued. This led the A's to pursue players who were relatively low paid but had a track record of success on the statistics that the A's judged valuable.

Long-time baseball people had and continue to have major objections to this approach. But it's hard to dispute the ability of the A's to win many games while always having one of the smallest payrolls in baseball. Moreover, people who did not really study the approach or who misunderstood the approach would often present a caricature of what the A's were doing. Understanding only pieces of the resulting strategy of the A's in the early 2000s and not the methodology and approach, detractors would say the strategy of the A's was to "get a bunch of fat guys who walk a lot, strike out a lot, and hit a lot of home runs." This is incorrect on many levels. There is no preference for "fat guys" (who likely don't steal many bases), just the recognition that stolen bases don't help you win that much. Getting walked a lot is not necessarily good in itself, except that it allows you to maintain a high on-base percentage without needing a high batting average. And the high batting average is overvalued in the market. Striking out is not good, but it's not necessarily something to dwell on if the player is doing the right things in his at-bats where he doesn't strike out. Finally, hitting home runs is a good thing for sure because at worst it guarantees to score a run.

Besides misrepresenting the strategy of the A's in those years, the more significant error on the part of detractors was to miss the larger point. Billy Beane did not have a preconceived notion about how to build the team and what qualities to value in a player. If the analysis would have proved that the ability to steal bases was undervalued by the market, we bet the A's would have pursued those players. Beane's decision was to make the best attempt to understand the statistics, understand the ability of those statistics to predict team wins, and then assemble a team that would allow the A's to win the most games within the budgetary constraints. As years went by and more teams learned from the success of the A's, the strategy had to change because the market was no longer undervaluing the player attributes that it had previously.

Moneyball makes a great case study for analytics because it discusses at length the use of descriptive analytics and predictive analytics by the A's to gain an edge over the team's more well-heeled competitors. What is not clear from the book is to what extent the A's used prescriptive analytics (that is, optimization) in making roster decisions. Thinking through how optimization could apply to this problem provides a nice introduction to optimization.

Certainly, Billy Beane and his team would weigh the predicted value of a player versus the salary they would pay him and decide which players were worth pursuing. There is no question that they were looking to optimize (that is, maximize) the number of wins of the team, but it is not clear if they used formal mathematics in this process.

To provide an example of an optimization-based approach to this problem, assume that we make a leap and use a single statistic to predict the number of wins a player will contribute to the team. The authors of this book are not Sabermetricians, so apologies if we pick the wrong one, but for the sake of argument, we'll choose VORP (value over replacement player). The idea behind VORP is to judge how valuable a given player is compared to a "replacement player." There is obviously a lot of debate and ambiguity about what level of achievement this fictitious replacement player would provide. But the simplified version of the idea is to say, for example, the achievement a team could expect from a bench player or a Minor League player that could be called upon to fill the Major League position. A starting player with a high VORP implies that the team would lose quite a few more games if this player were lost and the "replacement player" were used instead. Say, for example, that your starting third baseman had a VORP of 8. This would imply that the predicted reduction in wins over a season would be 8 games if he were injured and replaced in the lineup by an average replacement (someone from the bench or previously in the Minor League). So, let's suppose we agreed that the best way to maximize our expected wins would be to maximize the total VORP of our starting roster of players.

Our optimization problem is about maximizing VORP, subject to our budget constraint. If each player we are considering has a known VORP and salary, then this problem is exactly like a well-known optimization problem called the *knapsack problem*. To understand the knapsack problem, you can think about the origin of the name the problem was given. Suppose you are going on a long trip and need to decide what to put into your knapsack (or backpack). The knapsack can hold only so much weight. You have a variety of items you can put into the knapsack, each with a different weight and different value—like clothes, sleeping bag, shaving cream, flashlight, and so on. If you put in your large sleeping bag, you may not have room for the flashlight. Deciding what combination of items will produce the most value and fit into the sack is an interesting optimization problem. This problem pops up in a variety of places.

In our baseball roster version of the knapsack problem, the weight of each player is his salary, his VORP is his value, and the size of the knapsack is the budget limit. A naïve approach to solving this problem could be to use a "greedy" heuristic. One could imagine taking the VORP of each player and dividing it by his salary to get the "bang for your buck" of each player. You could then sort the players in decreasing order of VORP/salary and add players until you no longer have enough budget to accommodate the next player on the list; you would then skip players until you reached one that could still fit into your budget, and you would be done when there were no longer any players who could fit into your budget. The problem with this, and many other greedy approaches, is that they can paint you into a corner. For example, in 2002 (the primary subject of *Moneyball*), the team had a payroll of approximately $40 million. The best player in baseball at that time was arguably Alex Rodriguez, whose salary was approximately$22 million. Even if Alex Rodriguez's VORP to salary ratio was the highest in the league (it wasn't), it still would be very unlikely that the best strategy of the A's would have been to pursue Alex Rodriguez, for if they had been successful, they would have tied up over 50% of their payroll in one player and would have had very little flexibility to add additional high-value players to the roster. A similar problem in the opposite direction could occur if all the very high VORP/salary players achieved this, based not on having a high VORP but on having a low salary. In this case, a greedy approach might result in filling the entire roster with very low-priced players, leaving unused budget. In this case, your VORP/salary would be maximized but your total VORP would not, which is your actual objective.

In truth, the problem the A's would face would not be a pure knapsack problem because of some additional constraints that would certainly need to be considered. Namely, baseball players are far from completely interchangeable because they play specific positions. A solution in which we filled the entire roster with players who play shortstop would not be a very good solution. Instead, we would want to fill our roster with an appropriate number of players at each position. For simplicity, let's assume that we are concerned only with our starting lineup. This would mean that we need one of each position player (C, 1B, 2B, SS, 3B, LF, CF, RF) and a collection of pitchers (this can vary by team, but let's assume five starting pitchers, and six relief pitchers). We are left with the following optimization problem:

Target:	Maximize VORP
Limits:	Total budget
	Number of players assigned to each position exceeds the number desired
	Each player is assigned to only one position
Choices:	Which players to include in the roster and which position he will be assigned to
Data:	VORP of every player
	Salary of every player
	Eligible positions of every player

(Note we will provide a more detailed treatment of the targets, limits, choices + data construct in Chapter 8, "Prescriptive Analytics (aka Optimization).")

This problem can easily be formulated as a mixed integer program (more on this in Chapter 8). Given that the number of Major League baseball players who would be realistic options would be less than a couple thousand, a commercial mathematical solver would likely solve this problem in a matter of seconds. The result, given the assumptions stated, would provide the best possible roster the team could assemble to maximize its expected wins.

Of course, this would be very unlikely to produce a realistic roster in that we have not yet considered the players who were already under contract on the A's, which players were free agents (meaning they could sign with any team), and which players were under contract with other teams. Nevertheless, the solution could be instructive as a sort of aspirational roster to aim at in the future.

The good news is that we can supplement the optimization model to better incorporate the more realistic limitations of our current roster as well as the status of other players (free agent or under contract with another team). The most conservative such model would assume that we are locked into all the players we have under contract, and we may pursue only players who are free agents. In this case, even if the majority of the roster is set, there still may be many choices that can be made because there are many free-agent choices, and many players can play multiple positions. So, there are likely many combinations of players and position assignments that are possible. More likely, we would not assume that all the players currently under contract are locked on our roster (we could trade some to other teams); likewise, all players under contract with another team are not off limits (we could trade for those players). So, it would be the job of the modeling team, under direction from management, to lock in the appropriate players and make available the appropriate players for other teams. We might also imagine setting a limit on how many total players would be traded away and traded for because trading among teams is tricky.

As you can see from this discussion on how the A's could use optimization to make better decisions, there are a lot of factors to consider. What might not be so obvious is that you should think about the optimization model being a decision support tool, not necessarily the decision-making tool. That is, don't think of loading all the factors that impact a decision into the optimization model, clicking the Run button, and then having the program come up with the single best answer. Instead, think of the optimization model as allowing you to explore many different solutions, and each time you click the Run button, it does the hard work of sorting through all the possibilities and gives you the best result for that given set of input values. You may get a great solution that involves trading seven of your players. In this case, you might want to change the model so that it gives you the best solution that involves at most two trades. Maybe the projected win difference between the scenarios is only three games, so it is worth it.

It very well may be that the A's as well as other professional sports teams already use mathematical optimization to help guide their budget-limited roster decisions. To date, however, we have seen primarily a focus on descriptive and predictive analytics in the context of professional sports, and we have not seen anything written that would indicate much use of the field of prescriptive analytics. In Chapter 8 we explore prescriptive analytics in more detail.

8

Prescriptive Analytics (aka Optimization)

Prescriptive analytics takes data as an input and prescribes a solution. That is, prescriptive analytics helps you decide what to do. It helps you make a decision by sorting through all the different options you have and returns the best course of action to achieve the stated goals. You can set up a prescriptive analytics system so that it is embedded into an operational system and the results are automatically executed without human intervention, or you can use it strategically so that the results are presented for a decision maker to review and okay before they are implemented. *Prescriptive analytics* is basically synonymous with *mathematical optimization*, and we use these terms interchangeably.

It is our experience that prescriptive analytics is somewhat of the neglected child of the analytics family (like Jan in the Brady Brunch for those old enough to remember). Anecdotally, if you look for posted jobs on LinkedIn with the keyword *analytics*, you will find many employers looking for skills in statistics, probability, business intelligence, and so on, but very few that specifically ask about optimization skills. This is unfortunate because optimization, used properly on the right problems, has been proven time and again to provide tremendous benefits to organizations of all shapes, sizes, and purposes. Two documented examples in the following paragraphs help illustrate this point.

We suspect that optimization is neglected because people think that if they can do a great job of just understanding a problem (descriptive analytics) or of predicting how the future will play out (predictive analytics), then they will have the wisdom to choose the best possible solution to the problem. This may in fact be true when there is a very small range of choices, but in many situations the number of choices is very large (often surprisingly so because of the combinatorial nature). Let's go back to the problem of the A's filling their roster (refer to Chapter 7, "Case Study: *Moneyball* and Optimization"). Suppose the A's had 25 possible players to use as starting pitchers and 5 slots. It would probably surprise many people to know that this adds up to more than 53,000 possible combinations to consider. This is actually a tiny number for a mathematical optimization program to consider, but it still exceeds (by a wide margin) what a human being can digest. The number of choices is also harder to sort through when you have a limited amount of resources to invest.

In many cases, we see the need for all three types of analytics: descriptive to understand the data and what has happened historically, predictive to make scientific estimates of what could happen in the future for a variety of different situations, and prescriptive to sort through the thousands or millions of possible choices and move toward the ones that best meet the goals. Unfortunately, it seems that many see the field of analytics as being primarily about the first two.

Our recommendation is that it's great to understand the data, and it's better still to be able to make good predictions about the future, but you shouldn't stop there. Leveraging prescriptive analytics to enhance decision making can provide the final competitive edge.

If you're new to optimization, you may not realize that you often interact with it in your daily life. For example, when you get directions in Google, Bing, Apple, or some other mapping service, you are using optimization. You input the starting and ending points, and the underlying mapping service looks at all the roads and tells you which route you should take. The mapping service also provides descriptive analytics by drawing the map and route. It may use predictive analytics to predict drive times and traffic. But it definitely needs prescriptive analytics to recommend the best route between your starting and ending points.

Mathematical optimization is effective and widely used in many different businesses and organizations, from small to large. In finance, it is used to help build better portfolios. In manufacturing, it is used to schedule factories. In transportation, it is used to pack and route trucks. Television stations use mathematical optimization to decide which time slots to sell to advertisers, what they should charge, and in what sequence the ads should be placed. Major League baseball uses optimization to produce schedules that help minimize travel, maximize revenue, and adhere to logical constraints (for example, the Chicago White Sox and the Chicago Cubs should not both play home games on the same night).

The knapsack problem, like the one we suggested for the A's in Chapter 7, has many different applications.

There is now a lot of talk about a staffed mission to Mars. If you think about the vessel that will carry the astronauts as the knapsack, you have this same problem. The mission planners need to carefully decide how much food, water, medicine, and scientific equipment should be placed in the vessel to maximize the chance of success (and survival).

For a knapsack example in finance, think of a CFO having a limited amount of money to invest in projects for the year and having a lot of different projects with different investment amounts and different returns. In this case, the optimization problem is to maximize the expected total return with the limited funds. The limited funds are the equivalent of the size of the knapsack.

For a marketing example of the knapsack problem, suppose you are VP of marketing for a major consumer packaged goods company. There are a variety of proposals on your desk regarding different potential marketing campaigns. Your analytics department has done a very thorough job of researching how similar campaigns have worked in the past, both for your company and others. Based on this, as well as a variety of demographic data, industry trends, and so on, the analytics team has predicted the sales uplift expected from each of these campaigns. Now, given that each campaign has a different cost and expected sales uplift and given that you have a limited budget, how do you choose which campaigns to pursue?

For a construction knapsack example, suppose you are a home builder. You know buyers will pay more for certain upgrades (granite vanity tops, high-end appliances, fireplaces, and so on). You have estimates for the expected increase in sales price from each possible upgrade as well as the increased construction costs. You have a strict construction budget. Which upgrades do you choose to add? Again, this is a knapsack problem.

These are all examples of the knapsack problem, which only scratches the surface of the total number of applications for optimization.

The next example demonstrates the power and value of optimization in a completely different context. We suspect that almost everyone who reads this book has been in some way affected by cancer in their lives, either directly or through an immediate or extended family member. It may surprise readers to learn of the success of mathematical optimization in helping with cancer treatment. Specifically, a team of researchers at the Memorial Sloan-Kettering Cancer Center and Georgia Tech won the Edelman Award in 2007, an annual prize awarded by INFORMS (Institute for Operations Research and the Management Sciences).[1] The team won the award for using optimization to develop an improved process around brachytherapy, which is a treatment that involves placing radiation "seeds" within a tumor. The team developed a way to optimize the number and placement of these seeds in the treatment of prostate cancer. This method has had multiple terrific outcomes. First and foremost, it has improved the patient's experience by providing a less invasive procedure (15% fewer needles were needed, as well as 25–30% fewer seeds) and postoperative complications were reduced by 45–60% because the improved placement had less impact on healthy areas around the tumors. Second, this newly optimized process eliminated the need for doctors to spend many hours prior to surgery manually planning where and how many seeds they intended to place, and it also shortened the actual surgery time. In addition to the deeply personal saving of lives and improved quality of life, the financial savings from this efficiency improvement are estimated at several hundred million dollars a year in the United States.

Another example is documented in Steve Sashihara's *The Optimization Edge*. This example demonstrates that you don't need to be a huge corporation to derive the benefits of optimization. Jan de Wit Company is a Brazilian company and a large wholesale producer of bulb flowers. In the early 1990s, it began growing a small number of lilies in addition to its regular business. After about 10 years, the company had about 30 people working in this area of the business, which was very complex due to the need to buy bulbs from Dutch suppliers prior to when demand was known, grow different varieties in different greenhouses at different conditions, and so on. This complicated production planning process was all performed manually by the owner of the company. Spurred on by some optimization enthusiasts from the University of Sao Paulo, Jan de Wit started using an optimization program to plan production. In its first year of use, revenues increased 26%, and the contribution margin went up 32%.

The rest of the chapter will provide you with a deeper understanding of what optimization is and how it works. The optimization methodology will also provide you with a structured framework for thinking about solving problems.

What Is Optimization?

Much like the term *analytics*, the term *optimization* is used so frequently and in so many contexts that there is a danger that the term will become meaningless. In virtually any business context, a manager will be interested in improvement. When a project is undertaken to improve a business function, it is natural to say you are looking to *optimize* that function. We are not here to scold and tell people what words to use, but we do believe it is important to distinguish optimization as a discipline within analytics from the more casual use whereby you are simply saying you are trying to improve something.

Merriam-Webster defines *optimization* as "an act, process, or methodology of making something (as a design, system, or decision) as fully perfect, functional, or effective as possible; specifically: the mathematical procedures (as finding the maximum of a function) involved in this." As you can see, this definition says something stronger than simply to try to improve a "design, system, or decision"; instead, it specifies trying to make something "as fully perfect, functional, or effective as possible." In addition, it refers to that fact that mathematical procedures are involved.

Optimization is the primary discipline we discuss in the subcategory of prescriptive analytics. Recall that with prescriptive analytics, you seek to provide guidance to decision making. Optimization seeks to employ an algorithm-based approach to decision making, to replace (as Steve Sashihara calls them in *The Optimization Edge*) the three *H*s: history, hunches, and hierarchy.

A construct that we have often used to explain optimization to those who have not studied the field is to introduce the acronym *TLC+D* (Targets, Limits, Choices + Data).[2] That is, optimization is concerned with achieving *targets*, while adhering to *limits*, by making *choices*. Critical to this process is the use of *data*. Combining this with the characteristic of using an algorithm-based approach, we define *optimization* as follows: "The use of mathematically based algorithms to prescribe decisions that maximize or minimize defined targets while satisfying all known constraints."

Optimization = Targets, Limits, Choices + Data

In this section we rigorously (but hopefully not tediously) explain the *TLC+D* construct. We do this for two reasons. First, for those who have no previous experience in optimization, we believe it is an excellent way to teach the major elements and gain understanding about what optimization is all about. Second, thinking about these problems in these terms can greatly help clarify and set boundaries about what problem you are trying to solve, what you are trying to achieve, and what key considerations you must take into account. Any promising optimization project can quickly be sunk if the team is not careful to understand and limit what the optimization is intended to do, what will be considered and what will not be considered, what decisions are fixed and which are open to be changed, and what information is available to support these decisions and what is not. Going through the exercise of discussing, debating, and defining each of the components of an optimization problem can go a long way toward clarifying the goals of the project, setting reasonable expectations, and capturing the key ingredients of success.

Targets

When evaluating any solution to a given problem—a good, terrible, or mediocre solution—it should be possible to calculate the value that the solution provides. We call this the *target*. Optimization algorithms are concerned with minimizing or maximizing defined targets.

Since we need to compare the targets in a systematic way (we need to figure out the best one), these targets must also be able to be expressed mathematically. The process of doing this can be a valuable exercise in itself as it can enforce some rigor and discipline in a decision maker's thought process. If a target cannot be expressed mathematically, this likely indicates that it is a bit too fuzzy in nature to truly optimize, and therefore the team should try to define it more concretely. Understanding and defining the appropriate targets for a given business problem is perhaps the most critical component of a well-designed optimization process. The reason this is so critical is because, in a word, optimization algorithms are *ruthless*. What do we mean by this? Well, most serious optimization problems will be solved using mathematical algorithms that have been programmed into software. These algorithms are designed to find solutions to a given problem that perform as well as possible on the given target—this is their sole purpose for being. The algorithms seek to improve the targets ruthlessly; there is no nuance, there are no "yeah, but's." Within the constraints given, the algorithm will do anything to improve the target by any amount, no matter how small.

For example, suppose you are managing a car rental franchise. Your fleet of cars represents a capital investment. You know, inherently, that in order to achieve a good return on your capital investment, you need to have your cars rented a high percentage of the time (that is, you need a high utilization of each vehicle). It is critical that you decide how many cars you should have in your fleet. Now suppose your niece has just come home from college for the summer, fresh from taking a mathematical optimization course: She offers to help you with this question by leveraging this powerful form of analytics she has just studied. She asks what your optimization target is (or she might say *objective function*, in her academic jargon). You tell her that you would like to maximize the utilization of your fleet. Great, she says, and goes off for a week to carefully construct a mathematical model to maximize fleet utilization. After a week, she returns very excited, with the excellent news that she has found a solution whereby you can expect a 100% utilization of your fleet. Shocked, you ask her how this is possible. She responds that this can be achieved by maintaining a fleet of one car.

In this example, the optimization algorithm did its job. Your niece asked it to maximize fleet optimization, and it did. Certainly no one can claim that it is possible to devise a solution where fleet optimization is higher than 100%, and it is not hard to believe that 100% utilization could be achieved if you have only a single car at a rental location (that is, it is believable that there would always be demand for at least one car). The problem here is that, while utilization is certainly a worthy metric to track for a car rental manager, at least in this case it is not appropriate as the primary target of an optimization model. Think of an optimization algorithm as an extremely skilled archer who, in the best case, can be guaranteed to hit the bull's eye but at the very least can be trusted to do whatever he can to get as near the bull's eye as he can. Therefore, it is extremely important to place the target in the right place. Putting the target in the wrong place will likely produce a solution that will not improve—and may very likely make worse—the results you want to achieve. To use another analogy, the algorithm in Google Maps to create driving directions will almost always provide a very good route. However, if you have entered the destination as Springfield, Missouri, when you intended Springfield, Illinois, the resulting route will not have much practical use.

The targets of an optimization often align with specific key performance indicators (KPIs) at a company. KPIs are specific metrics that various people at different levels of management look at to get a quick idea of how a company is doing. The goal of a KPI is to give a manager an indication of whether the organization is performing well or poorly, based on where it is on the KPIs versus the targets for those measures. As in the case of the car rental manager, it may be that a valid KPI (for example, utilization) just doesn't serve well as an optimization target. But it may also be the case that certain KPIs work well for a segment of the business but may conflict with other business segments or and can distract from the primary targets or goals of an organization. Classic examples of this occur when different parts of an organization have different KPIs, and these KPIs conflict with one another. Consider a manufacturing company where the manufacturing manager has a KPI based on manufacturing output, while the warehouse manager has a KPI based on inventory reduction. The manufacturing manager has incentive to produce large batches of product at a time. This will reduce the need for changeovers and maximize the total output he can achieve. Unfortunately, this policy goes directly against the goals of the warehouse manager in that he will receive large quantities of product that will end up sitting in inventory for a long time and causing him to perform poorly on his inventory KPI. In this case, if your optimization problem were to optimize the production schedule, you would have at least two targets that should be considered: maximize output and minimize inventory.

In nearly all textbooks on mathematical optimization, problems are formulated using a single target. (In mathematical jargon, the target is known as an *objective function*.) However, as you have seen in the example of the manufacturing company, most real-world problems have multiple targets for optimization. The following are some examples:

- Investors would like to maximize expected return and minimize risk.

- Retailers want to maximize in-store availability and minimize obsolete or discounted stock.

- Sports teams would like to minimize payroll while maximizing wins.

- Airlines would like to maximize the number of passengers on an aircraft while also maximizing customer satisfaction.

Optimization Problems with Multiple Targets

There are several possible ways to consider multiple targets in the optimization process. In some situations, it may be possible to give weights to the different objectives and create a combined objective by summing the weighted pieces. For example, suppose you know that target 1 is twice as important as target 2, which is twice as important as target 3. In this case, you might create a combined target that equals [target 1 + 0.5 × target 2 + 0.25 × target 3]. You can then optimize your combined target to find a solution that properly weighs the three different targets you have set.

Although this seems like a simple and intuitive solution to optimizing for multiple targets, there are several problems with this approach. First, it may be that the different targets use different units of measure, which makes weighing them very difficult. Suppose an airline has targets for profit (measured in dollars), on-time arrivals (measured as a percentage), and customer satisfaction (measured as an average survey score). Even in the "easiest" situation, where the CEO has decided that these three targets deserve equal weight, what does that mean? You would need to somehow equate $1 in profit with a percentage of on-time arrivals and customer satisfaction score...and this is no easy feat. Even if the unit-of-measure obstacle can be overcome, you are still left with the arbitrary nature of setting weights. We have all seen situations where weights are given to different targets, which are then combined to a single value. Take, for example, the annual *US News* ranking of universities. A variety of targets are listed (SAT scores, alumni giving, class size, and so on). The magazine gives a weight to each, sums up the weighted score, and orders the results. This system misses the fact that there is no objective way to set the weights, and setting the weights differently could produce totally different results. Who is to say which is best?

Luckily, there are several more promising ways to use optimization with multiple targets. One common way is to choose what is truly the "primary" objective but to set limits, or constraints, on the other objectives so that the resulting solution performs at an acceptable level. For example, suppose an airline CEO sets a target of maximizing profit but states that on-time arrivals must be at least 80%, and customer satisfaction must have an average score of at least 4 out of 5. A good practice would then be to perform sensitivity analysis with changes to these constraints. (For example, how much could profit improve if you reduced the on-time arrival target to 75%? What if you increased it to 82%?) Another approach might be to use what we call *hierarchical optimization*. With this approach, you run successive optimization runs, one target at a time, in order from highest-priority objective to lowest-priority objective, using the result at each step to create a constraint at the next step. To illustrate, suppose the CEO decides that profit is the first objective, customer satisfaction is second, and on-time arrival is third. It is possible to set up an optimization routine whereby you do the following:

- Find the solution that provides the maximum total profit.

- Find the solution that maximizes the customer satisfaction score but does not reduce profit from step 1 by more than $x\%$.

- Find the solution that maximizes on-time arrivals but does not reduce profit from step 1 by more than $x\%$ and does not reduce customer satisfaction from step 2 by more than $y\%$.

Of course, hierarchical optimization can also be improved by performing sensitivity analysis. You might consider exploring the solutions when the sequence of targets is changed or when you vary the choice of x and y.

Finally, perhaps the ideal way of addressing optimization problems with multiple targets is to use a technique we call *multi-objective optimization*. In this case, instead of focusing on producing a single solution, the focus is to produce a series of alternative solutions that are considered *Pareto optimal*. A Pareto optimal solution is one where you cannot improve one target without simultaneously making another target worse. For example, suppose the airline example has a possible solution with $100 million in profit, an on-time arrival record of 78%, and an average customer satisfaction score or 4.1. If there are any possible solutions where one target improves (for example, profit goes up) but the other targets do not get worse (that is, on-time arrival and customer satisfaction are at least as 78% and 4.1, respectively), then the solution would not be Pareto optimal. Pareto optimal solutions are ones where there is no free lunch...you can't do better with one of your goals without another goal suffering. Multi-objective optimization algorithms provide techniques to generate a series of Pareto optimal solutions. This method can be extremely valuable to you because it helps you truly explore the inherent trade-offs between your targets and come to a well-informed decision of how you value the various targets and what solution you feel strikes the best balance. Given this, why ever consider the other methods? Primarily, the reason we would not recommend always using multi-objective optimization are technical in nature. First, most software packages and optimization solvers are not prepackaged to use this technique. Therefore, multi-objective optimization typically requires someone with a high skill level to create the algorithms. Second, the type of algorithms used can quickly become extremely computationally complex as the number of targets increases. In our experience, these algorithms can perform very efficiently on many problems where two objectives are being considered, but going beyond two targets may be impractical. An additional benefit of working primarily with two targets in multi-objective analysis is that the resulting trade-off can be viewed graphically, as shown in Figure 8.1.

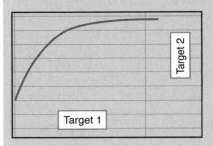

Figure 8.1 Sample Trade-Off of Two Targets

As you have seen in the previous discussion of descriptive analytics, being able to visualize results can deliver tremendous value in terms of the ability to comprehend the data and make good decisions.

Limits

As previously described, optimization algorithms are ruthless in seeking to minimize or maximize the stated targets of a problem. If real-world limits, or constraints, are not defined clearly in a way that an algorithm can consider, they will be ignored. For example, a popular application of optimization is in the area of revenue management (which we discuss at length in Chapter 9, "Revenue Management"). Most typically associated with the travel industry (for example, airlines, hotels, car rentals, and so on), revenue management involves creating a dynamic pricing strategy to maximize the revenue of the company. Companies that benefit from revenue management can expect to sell more of their product (seats, rooms, rentals, and so on) if their prices are lower. Projecting the price sensitivity (expected unit sales versus price) is the job of predictive analytics. Given this relationship, it is the job of optimization to guide the best pricing strategy in order to maximize revenue (price × units sold).

This might seem a very simple problem. Suppose you have a predicted relationship between units sold and price like the one shown in Figure 8.2.

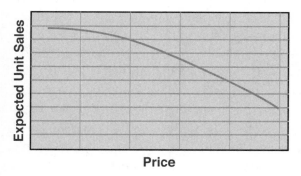

Figure 8.2 Sample Demand Curve

As prices rise, you expect to sell less. In order to calculate the revenue, you simply need to calculate the area of the rectangle to the Southeast of a point on the curve, as shown in Figure 8.3.

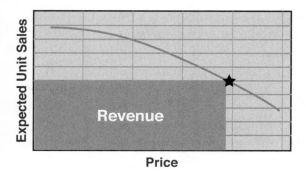

Figure 8.3 Sample Demand Curve with Revenue Calculation

The optimization problem at this point therefore boils down to the decision of finding the point on the curve that results in the largest rectangle. You can imagine this being done fairly easily by simply exploring a variety of points on the curve in Excel. Alternatively, if this curve is being defined by a well-understood mathematical formula, you can simply find the revenue-maximizing point by writing the revenue formula (price × unit sales) and finding the inflection point by using calculus (taking the first derivative and solving for the price, such that the result is 0).

In reality, however, this simplistic method to maximize revenue is very likely to miss some critical aspects of the problem and therefore provide recommendations that you cannot act on. To provide a very simple example, suppose you are in the situation of setting the price for hotel rooms. It may be that the revenue-maximizing decision would appear to be to sell 300 rooms at $200 per night. This is not a helpful recommendation if the hotel has only 225 rooms.

Representing the limit on the number of rooms in the hotel seems obvious and can very easily be represented as a mathematical constraint in the optimization model. Most problems, including this one, will likely have quite a few more constraints that are less obvious but that are nonetheless important to consider. Taking the next step in the examples of airlines, hotels, and car rentals, you would have capacity constraints for subcategories as well (first-class or coach; suites, king rooms, or double rooms; and full-size, mid-size, or compact cars). You would also need to create constraints reflecting the ability to substitute (that is, it is possible to offer a mid-size car to someone who requested a compact if you are out of compact cars, but you would not want to offer a compact to someone who requested a mid-size—or at least you would be quite reluctant to do so). Finally, in these particular problems, you would need to ensure that the constraints reflect the reality of the time dimension. Specifically, you cannot sell a seat on a flight, a night at a hotel, or a car rental in the past. Once the flight has taken off, or the day has passed, those items are no longer available for sale. This may seem blindingly obvious, but unless you build this aspect into your optimization problem and have proper constraints to reflect this reality, an optimization algorithm will continue to try to increase revenue by selling seats on last week's flight.

Understanding and specifying the limits is critical to the success of an optimization model. In the process of implementing an optimization project for a business decision, it is wise to spend time at the beginning trying to capture these limits to feed the model. In our experience, however, it is nearly impossible to capture all of these up front. Invariably there are limits that exist that don't come immediately to mind and are only discovered when you start reviewing recommended solutions and someone with deep knowledge of the business says, "We can't (or wouldn't) do that because of X." At that point, it is up to the team to incorporate this additional limit into the model.

For example, we worked on a project with a client where the goal was to optimize the assignment of customer demand locations to the service centers. In some initial solutions, there were instances in which a customer was assigned to a service center in another city, even though there was a service center in the same city as the customer. From an optimization point of view, this made sense because the capacity of the service centers was limited, and if we assigned this customer to its "home" service center, several other customers would need to be reassigned, and the total cost of operations would increase. Despite the cost efficiency of this possibility, knowledgeable managers from the company stated unequivocally that this should not happen in practice as it would be too unintuitive for both the customers and the operations team. As a result, we added this limit—that if there was service center in the same city as the customer, the customer must be assigned to that service center—to the optimization model. The resulting solutions aligned with expectations of management.

As important as it is to state all the important limits to be applied to the problem, it is also important to understand the difference between "hard" constraints and "soft" constraints. A hard constraint is one that we should not violate under any circumstances. If a solution is offered that violates any of the hard constraints, the solution should be considered *infeasible*. On the other hand, soft constraints can be thought of as constraints that you would like adhere to, but there is possibly some wiggle room. While you would prefer a solution that does not violate any constraints, a solution that only violates soft constraints may still be possible to implement. Too often, we have seen companies treat all constraints as hard. This is often revealed when a variety of constraints are listed as hard constraints and the team discovers that these constraints were frequently violated in practice in the past. It can be a useful exercise to see what solutions are possible when all constraints are treated as hard (as a sort of aspirational state), but there two good reasons that the model should also be evaluated with true soft constraints treated as soft. First, it is certainly possible that there are no feasible solutions to the problem with all hard constraints, which although useful to know, leaves you stuck with no recommended solution. Second, even if feasible solutions exist that satisfy all constraints, there may be solutions that perform much better in a number of ways that only violate the soft constraints in a relatively benign way.

To illustrate the difference between hard and soft constraints, let's look at the classic optimization problem of vehicle routing. Say that you have a number of deliveries you need to make to different customers in different locations. These deliveries are of various sizes, and the customers have different expectations on when they would like the deliveries to occur (for example, 9–11 a.m. on Monday). Other key parameters are the capacities of the trucks, the fixed costs, and cost per mile of the trucks and the drive times between locations. Given these parameters, optimization can be used to find a solution (truck routes and schedules) to make deliveries at the lowest cost possible. Two limits should be obvious. First is the capacity of the truck. You can't put deliveries on the truck such that the total size of the deliveries exceeds the capacity of the truck. Second, a limit on the solution can be created so that the delivery time to each customer is within the customer's desired window. You might start your analysis by treating both types of limits as hard constraints. When doing so, it is certainly possible that there are no available solutions that satisfy both limits. A second possibility is that in order to meet some customer's delivery window, you need to send trucks with very low utilization. Imagine a situation in which a truck needs to be sent to a customer with only that customer's shipment, and only 10% of the truck is full. Clearly this seems less than ideal. By now you are probably anticipating that another possibility is to treat the time window limits as soft constraints. Perhaps this results in a solution where that low-utilized (10% full) truck is eliminated, but instead you will be 10 minutes late in delivering to that customer. This may very well be an attractive trade-off. These soft constraints can often be implemented by specifying penalties for violation of the constraint. In this example, you might apply a cost ($) per minute for delivery outside the desired time window. The optimization algorithm could then consider this cost versus the costs involved in routing the trucks (number of trucks and miles driven). In some cases, companies may have actual costs that can be applied as penalties (for example, Walmart may contractually penalize its suppliers for missing delivery windows), while in other cases there may be a bit of art involved in setting these penalties. Commercial optimization solvers can make it relatively easy to implement soft constraints and do some experimentation with the penalties involved so that the resulting solutions reflect the proper trade-off between satisfying these constraints and overall solution quality.

Finally on this topic, it should be noted that it can be an effective strategy to "layer" a soft constraints with a hard one. In the previous example, suppose you wanted to penalize the solution $5 per minute a delivery was late. In addition, you might want to say that under no circumstances should a delivery by more than 90 minutes late. The optimization model can easily incorporate a soft constraint with the desired delivery window and associated penalty for violating the window, plus a second (hard) constraint that sets the last feasible delivery time as 90 minutes past the desired time.

Choices

Choices (*decision variables*, in math jargon) are the items for which you would like the optimization algorithms to provide recommendations. Quite simply, choices are the things you are trying to decide. In effect, optimization tries out different values for the choices, makes sure the values of those choices don't violate any of the limits you set up, calculates the target value for a set of choices, and then compares the target values for various different choices and picks the best.

Going back to the *Moneyball* example from Chapter 7, the A's had choices to make about which players would be on the roster and which position each player would be assigned to play. As is often the case, a subtle point here is that from a modeling perspective, only the second set of decisions is truly a decision variable. Assuming that every player is assigned a position, if you decide the assignments of players to positions, you have implicitly decided the roster as well (that is, the roster is simply the collection of players that have been assigned positions). In the earlier example about a rental car location, the choices presented to the optimization model were the number of vehicles of each class to carry. The number of choices that are available will largely inform you about whether it is worthwhile to apply sophisticated optimization modeling to a given problem. If there are only tens or hundreds of possible choices to make, then someone who is quite adept with spreadsheets can very realistically perform enumeration of those choices and find the best one. For example, suppose the A's were only concerned with filling 3 roster spots, and there were only 10 possible players to choose from to fill those spots. In this case, the number of combinations to consider would be $10!/(7! \times 3!) = 120$. It may actually be a bit tedious to set up a spreadsheet model to test all 120 possibilities (we are not aware of a simple Microsoft Excel function that would generate the list of combinations), but nevertheless it is certainly workable. On the other hand, if the A's have 10 roster spots to fill, and 100 players to choose from, the number of possible combinations in this case would be $100!/(90! \times 10!) = 17,310,309,456,440$. This spreadsheet exercise very quickly went from tedious to impossible. The power (and, frankly, magic!) of optimization algorithms is that they could very likely solve a problem this size in a matter of minutes in modern software.

In any problem you would like to solve by using optimization, identifying and documenting the choices available is clearly a critical element. In some cases, this will seem like a simple exercise. However, in many cases, there will be some ambiguity or additional questions that need to be answered to fully define the decision variables of the optimization model. For example, a sort of classic problem used in classroom settings to teach optimization modeling is the problem of nurse scheduling. (As an aside, in most university settings, there tend to be far more men studying mathematical optimization than women. A few years ago, we asked a female colleague and optimization expert why it was that nurse scheduling was used so often in the classroom. She deadpanned, "Men like nurses.") Basically, the nurse scheduling problem is a problem that involves a roster of nurses with different skills, pay rates, and availability, and you have a set of shifts in different areas that need to be filled and must assign nurses to shifts in order to minimize the total costs. In the simplest case, you can think of a choice that says, for each nurse, for each shift, "Does this nurse work this shift?" If you were formulating this model, this would be known as a *binary decision variable*, meaning it can take only two values (yes or no). Generally, a binary variable will be represented by a 0 or 1 in the model. Suppose, however, that you added a bit more complexity to this problem. Suppose it was possible (albeit not preferred and therefore penalized in some way) to have shifts that were not assigned a nurse. You would then need to add to your collection of choices (again as a binary decision variable) whether a given shift was filled or not. Another possibility is that you could add the option of hiring temporary staff to work some of the shifts. You might be limited to hiring temp staff for entire shifts, in which case you can think of a new choice of how many shifts should be filled by temporary employees. In this case, you have added an *integer decision variable* to the model. This simply means that the variable can take any integer value (for example, 0, 1, 2, ...). Finally, in this example, suppose that the temp staff is extremely flexible and is willing to work any amount of time (not just entire shifts). In this case, the choice could be how many minutes (including fractional) to hire external staff. These choices could then be represented by *continuous decision variables*, which can take on any (nonnegative) value, including fractions.

Both targets and limits rely on your definition of choices. Mathematically, the target (objective function) will be written as a function of the choices (decision variables). In other words, it must always be possible to calculate the value of the target (for example, profit, cost, customer satisfaction, and so on) if you know the values of the decision variables. Similarly, the limits (constraints) will in some way limit the choices. Typically, limits are written as some sort of inequality (<=, =, or >=), with decision variables appearing on one or both sides of the equation.

Data

Optimization cannot provide benefits without appropriate data. Nearly everyone has heard the phrase "garbage in equals garbage out." This phrase became popular during the era of computer software's explosive growth to reflect the fact that the solution or output of a software program can only be as good as the information being input to the software. But in some sense, the problem is more serious than this for optimization models. If some process is clearly providing "garbage out," decision makers can simply discard that garbage and continue making decisions in other ways. The problem you face with optimization models is that they are most useful and worthwhile when confronted with very large and complex problems. In these cases, it may not be obvious or easy to discern that the recommended solution is "garbage," and because the purpose of optimization is to provide specific recommendations for action, you may discover the poor quality of the solution only after the fact.

When you look at the mathematical formulation of an optimization problem, all data looks identical (that is, it is simply numbers stored in tables, arrays, matrices, and so on). However, at least from a qualitative perspective, our view is that the data that is used in an optimization model falls into two categories. The first category is data that can be fed directly into a model and is in some meaningful sense certain. Any time you are working on an optimization problem for which there are operational decisions being made, it is likely that there will be a collection of this type of data available. Consider again the problem of deciding how to route trucks to provide a variety of deliveries. Much of the data that goes into this optimization model will be of this first type. That is, you will have a list of delivery points that need to be visited, the size of those shipments (weight, volume, and so on), the distance between points, the available trucks and their capacity, and so on. There may be some uncertainties, or approximations, in each of these data points, but the intent would be that the data is the data in this case, and it can be entered directly into the model.

The second category of data for optimization models is really geared toward what we would call the *parameterization* of the model. In this case, you are using data to make some sort of abstraction or approximation that will also feed the model. This may be a subtle point in many respects, but we believe there is a distinction, and it can be meaningful. When you work with parameters, there should be a greater recognition of the uncertainty or changing nature of these elements and, therefore, how you allow these values to affect the solution. In addition, these parameters of a model scream for sensitivity and what-if analysis (more on this later). Going back to the truck delivery problem, there are several pieces of data that are qualitatively different from the ones listed previously. When creating truck routes, it is important to know both the distance and the travel times between points. The distance, as implied above, is a number that you can pretty much enter directly in the model. Sure, the driver may decide to alter the route, or there may be small variations, but generally in the era of widespread GPS capabilities, the distance between locations is pretty well known or easily determined. On the other hand, the expected transit time between locations is much less certain. At best, you can make approximations. Times may vary significantly based on traffic, accidents, and so on. Various services (for example, Google) seem to be getting much better about including considerations such as traffic into their estimated travel times, but this remains a very difficult value to predict precisely, and it probably will remain so until self-driving cars are the norm. So if you are planning to optimize truck routes for the next day, you need some data on travel times, but the choice of what data to use is not clear. If you use the "best case" times or even the average times, you run the risk of making late deliveries if the driver runs into heavier traffic than expected. However, if you plan for the "worst case" times, then the drivers will almost certainly have idle time in their schedules as they are unlikely to encounter bad traffic for their entire route. Another element that will be important to the optimization problem is the costs you apply to the routes. To keep things simple, assume that you can apply a fixed cost for every truck you use plus a cost per mile a truck is driven plus a cost per hour the truck (driver) is used. This cost structure is rather typical for a company operating its own fleet of vehicles. These cost parameters will be important to the model because they imply a trade-off between different possible solutions. Imagine a solution that creates very compact, efficient routes but uses lots of trucks to do so. This solution would incur high fixed costs but relatively small variable (mileage cost plus hourly cost). On the other hand, a different solution might use few trucks, but the routes might be much less efficient (more miles driven and time used as the trucks move between points far from each other without much cargo in the trailer). The data you enter for costs (fixed, per-mile, per-hour) will obviously be very important as they will drive the solution that the algorithm produces. However, it will probably also be clear to those familiar with trucking (or really anyone familiar with cost accounting in an organization) that how you arrive at the costs to use in your model will not be an exact science. In this example, the costs you use will involve many factors, such as purchase or lease costs of the trucks, depreciation, ongoing maintenance, salary and hourly wages of the drivers, and fuel costs and fuel efficiency. Each of these factors might have a level of uncertainty, approximation, or assumption necessary to arrive at a number to use in the model. Moreover, these values may change over time and location (fuel prices) or by other factors that are difficult to consistently measure (for example, mileage may depend on the tire inflation of the trucks). These numbers will never be perfect because, in reality, there is no such thing as perfect in these situations. Despite these uncertainties, you should remember that the data is a means to an end, not an end in itself. The goal is to arrive at solutions that work for the problem you are trying to solve. Therefore, it is important to recognize data that has a degree of uncertainty, try to understand the level of uncertainty, and, most importantly, understand how these parameters impact the solution. This will help inform your analysis when you may need to revisit you data assumptions or when sensitivity analysis might be advised in order to ensure a good and robust solution.

In the context of optimization models, you will see data attached to all three other elements. The target will almost certainly include data—that is, the target will be a mathematical function that combines the data with the choices of the model. Similarly, the limits will be expressed as mathematical inequalities that combine choices and data.

TLC+D in Action: Everyone Loves Pizza

To illustrate an example of formulating an optimization problem from start to finish, using the $TLC+D$ construct, we present a small problem example that we have used in the classroom. The example is loosely based on a possible problem for a local pizza shop. At this pizza shop, you may have a sit-down meal, you may order carryout or delivery, and you can buy precooked frozen pizzas out of the in-store freezer. To keep things simple, let's assume that the restaurant sells only two kinds of pizza: cheese and supreme. The freezer has limited space, and making the pizzas in the morning will require time in the oven and from the cooks, both of which are limited. The different types of pizza have different profit margins. Finally, assume that these pizzas are so delicious that demand for them is unlimited.

If it were the case that the more profitable pizza (in this case, supreme) also required less resources (oven and labor), the best strategy would be quite simple: simply fill the freezer with supreme pizza. However, supreme pizzas actually take longer to prepare (more vegetable chopping and so on) and also take longer in the oven to get the vegetables and meat cooked properly. Because the more profitable pizza requires more resources and the resources are limited, you are face a nontrivial question regarding how much of each type of pizza to place in the freezer in the morning.

Targets

For the pizza shop, the goal is to maximize profit. For any set of choices, you are going to sum up how much profit you would make. Then you'll pick the set of choices the leads to the most profit.

Limits

As mentioned earlier, there are limits on the storage capacity of the freezer, the amount of oven time available, and the number of labor hours available.

Choices

You have a choice of how many cheese pizzas and how many supreme pizzas to store in the freezer.

Data

The data you need to feed the optimization model is summarized in the following two tables:

Profit Margins & Capacities

Profit Margins and Resource Capacity

Profit margin per cheese pizza ($)	4
Profit margin per supreme pizza ($)	5
Freezer capacity (pizzas)	95
Maximum oven time (minutes)	3,000 (10 ovens for 5 hours)
Maximum preparation time (minutes)	840 (7 people for 2 hours)

Resource Consumption

	Freezer (pizza)	Oven (minutes)	Prep (minutes)
Cheese	1	30	6
Supreme	1	35	12

Mathematical Formulation

Now that you have described the $TLC+D$ for this problem, writing the mathematical formulation is pretty straightforward. You start by defining two decision variables (choices):

- Let C be the number of cheese pizzas to store in the freezer
- Let S be the number of supreme pizzas to store in the freezer

With these defined, you can write both the target (objective function) and limits (constraints) as a combination of the data and the choices:

Target: Maximize profit = $4C + 5S$

Limits:

Freezer capacity	$C + S <= 95$
Oven capacity	$30C + 35S <= 3000$
Prep capacity	$6C + 12S <= 840$

This formulation has been written in a form known as a linear program. Linear programs have been studied and deployed for years and are likely the most widely used optimization techniques that exist.

As shown here, you can think about testing different values for C and S, making sure those values don't violate limits, and then seeing the profit of each choice.

Next, we will discuss different types of algorithms for solving this type of problem. The pizza example is simple and could be done by hand, but real problems quickly become too difficult to do manually.

Types of Optimization Algorithms

For some problems, there are known algorithms that are guaranteed to provide absolutely the best solution; these are known as *exact algorithms*. Unfortunately, exact algorithms are not possible or practical for many problems. In some cases, there are algorithms that can guarantee a solution within a certain factor of the best possible solution; these are known as *approximation algorithms*. In other cases, it may be that an algorithm is employed that does not have any precise mathematical guarantee about the quality of the solution. In these cases, we rely on anecdotal proof of the performance of the algorithm compared to what had been achieved before the algorithm was employed. These types of algorithms are most often called *heuristics*.

Obviously, it would be preferable to have exact algorithms for all optimization problems (after all, why wouldn't you want a guarantee that you have the best solution?). So why do you ever use approximation algorithms or heuristics? The answer involves the field of complexity theory. Given our space constraints, we grossly oversimplify here (apologies to our graduate school professors for the following bit of hand waving). The basic explanation is that there are some problems for which there are known exact algorithms that are in some sense "fast," while there are many other problems where there are no known "fast" exact algorithms, and mathematicians don't know if it's possible to create "fast" exact algorithms. By "fast," we generally mean something along the lines of saying that the time an algorithm takes to find the optimal solution will not "blow up" as the problem size increases. For the more mathematically inclined, a fast algorithm is defined as being solvable in polynomial time. According to Wolfram MathWorld, "An algorithm is said to be solvable in polynomial time if the number of steps required to complete the algorithm for a given input is $O(n^k)$ for some nonnegative integer k, where n is the complexity of the input. Polynomial-time algorithms are said to be 'fast.'"

Exact Algorithms

As previously stated, an exact algorithm is an algorithm that is guaranteed to provide the optimal solution to a given problem. In some situations, an available exact algorithm will be enumeration. Depending on the type of problem and the range of values a solution may take, it may be (theoretically) possible to try every possible value. If you simply enumerate all possible solutions to a problem and then pick the one that minimizes or maximizes the target, you can be assured of having the best solution. There are several problems with this brute-force style of approach. First, many problems do not have a discrete number of possible solutions to try, so it is impossible to enumerate. Second, even if there is a finite number of possible solutions, this approach only works for relatively small problems. Unfortunately, with many real-world problems, the number of possible solutions is likely to exceed the number of atoms in the universe. Clearly, enumeration does not count as a "fast" algorithm. In these cases, you probably need a better plan. Luckily, mathematicians and operations researchers (among other disciplines) have been very active in the past 80 years or so, developing algorithms that can solve very large problems.

The first exposure that most people are likely to have to optimization (if they have had any) is through linear programming (or its close cousin mixed-integer programming). Linear programing (LP) provides a way of mathematically expressing an optimization problem in the language of algebra. Sometimes we are a bit sloppy in our language and describe LP as a way of solving a problem, but to be precise, it provides a way of *formulating* a problem. There are several well-studied exact algorithms for solving problems that have been formulated using LP, and there is very powerful commercial software to do just that. The earlier pizza problem is written using LP. The basic rules for modeling a problem using LP is that the objective function (target) and all of the constraints (limits) must only contain *linear* algebraic statements of the decision variables (choices), and the decision variables can take on linear values (are continuous). This means you can multiply the variables by constant factors, and you can add and subtract the variables (with constant factor multipliers), but you may not multiply the variables with one another, you may not raise the variables to an exponent, and you may not use mathematical functions like log, or min, or max, and so on. Mixed-integer programming (which commercial solvers are also very adept at handling) is just like LP, except that you can place an additional constraint that some (or all) of the decision variables must take on only integer values (not fractions). Look back at the pizza example, and you can see that it should probably actually use a mixed-integer programming formulation by simply adding the constraint that C and S (the number of cheese and supreme pizzas) must be integers.

With exact algorithms, there is mathematical proof when the optimal (very best) solution has been found. Depending on the type of algorithm, how this mathematical proof is presented can differ. For example, when deploying algorithms to solve linear programming problems, it is typical to prove optimality by comparing the solution value of the problem to the solution value of a related problem that you know must be a bound on the original problem. Specifically, for every linear programming problem that is a minimization problem, there is a corresponding maximization problem (or vice versa); this is known as a *dual problem*. It can be shown (that is, proven through mathematical logic) that the dual maximization problem can never have a value larger than a solution value for the original minimization problem. In addition, it can be proven that the optimal value of the original problem and the optimal value of the dual problem must be identical. Therefore, if you work to minimize the original problem while simultaneously maximizing the dual problem, you can conclude with certainty that you have the optimal solution when the two values meet.

In other cases, it may be possible to prove that an algorithm is exact (that is, guaranteed to deliver an optimal solution) by using other mathematical logic. Mathematicians might use a variety of techniques to prove that an algorithm is guaranteed to provide an optimal solution. Some techniques may involve algebraic or geometric properties, and others may be more logic based. For example, one method of proof for certain types of optimization algorithms (for example, scheduling algorithms) is what's known as a *proof by contradiction*. To use this technique, you start by assuming that the algorithm provided will *not* produce an optimal solution. The idea then is to show through a series of logical arguments that you must then arrive at a conclusion that can't possibly be true. Once this false statement is derived, you must conclude that your assumption was also false, and the algorithm must in fact produce an optimal solution (what you wanted to prove).

It is beyond the scope of this book to give any reasonable treatment of how mathematicians develop and prove the exact nature of their algorithms. The point here is that you need to understand is that people often use the terms *optimal* and *optimized* very loosely. It is important to know if the algorithms being used to find a solution are exact algorithms, whereby you will know that the best mathematical solution to a given problem has been found, or whether a person claiming to have an optimal answer is really only stating that it is "optimal" based on his or her intuition or anecdotal proof. If it is the latter, it may still be a valuable solution, but it should perhaps be given a greater degree of scrutiny to determine whether there is still opportunity for improvement.

Approximation Algorithms

As previously mentioned, there are many optimization problems for which there are no known fast (polynomial time) algorithms. Even more discouraging, the best mathematicians, operations researchers, and compute scientists do not know whether it is possible or impossible to create fast algorithms for these problems. There is an entire class of optimization problems for which these experts can prove, however, that if a fast algorithm exists for one of these problems, then a fast algorithm must exist for all of these problems. On the other hand, if it can be proven mathematically that it is impossible to design a fast algorithm for any of these problems, then you know that it is impossible to create a fast algorithm for any of these problems. The question of whether it is definitively possible or impossible to create fast algorithms for these problems is one of the great open questions in the fields of applied mathematics and computer science. In fact, in the year 2000, the Clay Mathematics Institute labeled this problem (among six others) as Millennium Prize Problems, each of which has a $1 million prize awarded to anyone who could solve them. To date, one of the seven problems has been solved, and it's not this one.

This short background is meant to further explain the motivation for the creation of approximation algorithms. The basic idea is a simple one: There are many problems for which you don't know how to guarantee that you can find the best solution in a speedy manner. For these problems, researchers came up with a different idea. The idea is to see if it's possible to develop algorithms that, while not guaranteeing the best solution, can guarantee a solution within a certain factor of the best solution and do so quickly.

Imagine this situation: You have a very difficult problem for which you would like to find an optimal solution. You present this problem to your analytics staff only to be informed that there are no known fast algorithms to solve this problem. The good news you are told, however, is that there is a fast algorithm that provides a solution that you can be assured is within 10% of the best possible solution. Given the complexity and importance of the problem, this may be a great solution.

If you are faced with a problem for which there is no known fast exact algorithm, a fast approximation algorithm may be a good option. Another situation where approximation algorithms may be appropriate is when a fast exact algorithm exists, but that algorithm is fast in theory but not in practice, while the approximation algorithm is fast in practice and has a good provable bound.

To differentiate theory from practice, we go back to our definition of fast algorithms. We consider an algorithm theoretically fast if the number of steps it takes to solve can be written as a polynomial expression of the size of the input. This means, for example, that we might say that the algorithm will take no more than I^2 steps, where I is the size of the input. An algorithm that takes no more than $I^{1,000}$ is also a polynomial time algorithm, however, and is therefore considered fast, in theory. You might think of being fast in theory as meaning that there is a finite amount of computing power that could be used to solve the problem, and if you had that much computing power, you could solve the problem quickly. You can think of an algorithm that is not even fast in theory, however, as requiring infinite computing power to solve in the worst case. In practice, however, the limits of computing power are based on the current state of technology, your budget for those resources, and so on. Therefore, an algorithm can be fast in theory, but when applied to real-world problems with real-world computers, the time needed to solve the problem is too long to be practical. (Suppose you are trying to create next week's schedule. If an exact algorithm takes two weeks to solve, it does not offer much practical value.) In these cases, it may be preferable to apply a good approximation algorithm rather than an exact algorithm. For example, suppose you had a problem that had an exact algorithm with a worst-case $I^{1,000}$ performance versus an approximation algorithm with a worst-case performance of I^2 and a solution guaranteed to be within 1% of optimal. Clearly, the latter might be preferable, depending on the specifics.

Heuristics

The term *heuristics* covers a very wide variety of techniques used to solve optimization problems. Unlike exact algorithms, a heuristic has no guarantee of finding the best solution to a problem. Unlike approximation algorithms, heuristics cannot even guarantee that you are within a certain factor of the best solution. So, why would you ever use heuristics? There are a few reasons:

- As discussed previously, you might be dealing with a problem for which there are no known fast exact algorithms, or the exact algorithms may be fast in theory but not in practice.

- It may also be the case that there are no known fast approximation algorithms or approximation algorithms that again are fast in theory but not in practice.

- It may also be that the guarantee an approximation algorithm provides is not practically useful. Certainly it is the goal of approximation algorithms to provide "tight" approximations, but if any guarantee can be made, then technically it is an approximation algorithm. So, for example, if you gave us a problem for which you wanted to minimize the cost, and we provided an algorithm that we told you was guaranteed to be within a factor of 1 million of the best solution, you would probably not consider our work very helpful.

The downside of heuristics, as we have discussed, is that in general we only have anecdotal evidence of the quality of the solutions they provide. But heuristics have some upsides:

- They are often easy to understand.
- They can often be applied to a wide range of problems.
- Good heuristics often provide "good" results—in the sense that they are shown to provide better answers than the previous process deployed to solve the problem or based on the evaluation of the subject matter experts within the decision process.
- They are (or should be) fast in practice.

Because basically any algorithm that does not qualify as either an exact algorithm or an approximation algorithm can be considered a heuristic, it would be impossible in this book to describe even a fraction of the heuristics out there. We would like, however, to describe a few of the most widely used types of heuristics. These deserve treatment here for two reasons. First, the reason these have been used so widely on real-world problems is that they have been shown to quickly and successfully provide good solutions to many different optimization problems. Second, we discuss them here in part to provide a word of caution: These algorithms have very scientific sounding names and indeed are developed by strong researchers who are constantly trying to improve their performance (both speed and solution quality). Because of this, the uninitiated could possibly be misled in overstating their mathematical precision. In the end, these are still heuristics, with no mathematically provable guarantee of the quality of their solution. So there is a certain amount of skepticism or at least realism on the table when these heuristics are used.

We could not possibly cover all the different types of heuristics deployed for optimization problems. Next we briefly describe three of the most widely used types of heuristics but do not provide a detailed account of the algorithms or their applications. You can find entire texts and courses on these topics. Our intent here is to introduce the topic and define some the vocabulary to try to demystify these scientific-sounding terms and set proper expectations about what they can deliver.

Genetic Algorithms

Genetic algorithms are a class of heuristics that attempts to use the ideas of biological evolution to solve optimization problems. The idea behind these algorithms is that they maintain a population of possible solutions and then, through a few techniques that mimic biology, you produce new generations of solutions that have evolved to become stronger. The primary way you do this is though "mating." In this case, you take different possible solutions and allow them to create "children" by combining characteristics of their parents. To create a child, you take a portion of Solution A and a portion of Solution B and then combine those to create a Child Solution C. There are a couple of obvious advantages in this mating process versus a biological one. First, you are not restricted to having only two parents for a child. You may decide it's advantageous for many parent solutions to be combined for a child (think of a mad scientist who wanted to take DNA from Michael Jordan, Steven Hawking, Charlize Theron, and Steffi Graf to create a genius, beautiful, world-class athlete). It's also possible to create many, many children without having to wait months. Generally, so that the population of solutions will remain reasonably sized, at each step the algorithm will only keep solutions that have reasonably strong results. The other technique that occurs as the algorithm unfolds is mutation. Mutation in this context involves picking a random solution and randomly changing some portion of the solution. This again mimics biological evolution, and a certain amount of random mutation can often improve the solution quality, just as a certain amount of random mutation of DNA has had positive effects on human ability. Genetic algorithms typically have logic to determine which solutions in each generation will survive to the next generation, which will be selected to mate, how much cross-over will occur during mating, how much mutation will occur, and finally when the algorithm will terminate. Just as biological species evolve to better their chances of survival, genetic algorithms are meant to allow solutions to evolve that best meet the stated objective of a given problem.

Simulated Annealing Algorithms

Simulated annealing algorithms get their name from their similarity to a thermodynamic process related to heating and cooling metals in a pattern that results in a "slow" cool of the metal that produces fewer defects in the final product. The basic idea in deploying these heuristics to an optimization problem is that you start with an initial solution and allow that solution to "jump" around significantly at first (meaning you try other solutions with significantly different characteristics), hoping to find better "neighborhoods" of solutions. But then over time, you are doing a slow cool, where the solution jumps are much smaller or less likely, tending to stay in their own "neighborhood" (solutions that are very similar) and finding the very best solution in that neighborhood. Remember, these algorithms have no guarantee of finding an optimal solution, but this is typically not a realistic goal when they are deployed. A simulated annealing algorithm aims to find a good solution in a reasonable time, and these algorithms have anecdotally been proven to do so in many cases.

Tabu Search Algorithms

Tabu search algorithms belong to a class of heuristics known as *local search* (algorithms that take a solution to a problem and look "locally" to other solutions to see if improvement is possible). Many local search heuristics essentially take a "greedy" approach in that they look for other local solutions and basically move to a new solution that shows the greatest marginal improvement. Tabu search algorithms add to this by keeping track of certain areas of the solution space that are in some way "bad" areas (for example, solution quality is poor or constraints are violated). These areas are labeled as "tabu," and as the algorithm proceeds, logic is used to avoid those areas as new solutions are explored.

What-if Analysis

One of the most useful aspects of applying mathematical modeling and optimization algorithms to a problem is that it very often allows for the possibility of doing significant what-if analysis. That is, once a problem has been formulated and an algorithm is working, it is typically very easy to test different ideas, modify data elements, and ask different questions and see how the solution may change. As mentioned before, many people's most frequent exposure to optimization is through the generation of driving directions from various GPS providers. When asking for directions, you often can enter different criteria, such as "Avoid Tolls," "Shortest Distance," or "Fastest." These are simple forms of what-if analysis. This type of analysis is possible if the problem is set up properly, with an algorithm ready to solve it.

Some what-if analysis could also be considered sensitivity analysis. Sensitivity analysis generally entails making perturbations around certain data elements of the problem and seeing how greatly the solution changes. We have used the example of optimizing the routing of delivery trucks on several occasions. You might be interested in understanding how much the solution would change if you assumed that it takes 30 minutes to unload at a customer site rather than the current assumption of 20 minutes. How would that change the schedule and routes? How much additional cost would result? Would it require additional trucks? Certain types of optimization algorithms may provide some automatic information about sensitivity without requiring you to rerun the algorithm. In these cases, you might be given some information on how the solution value will change "on the margins" for certain data or constraints. Nevertheless, this would typically only give you information on how the solution value would change (for example, the cost or profit), but it would not give you guidance on how or when you might actually change your plan. Therefore, it is very typical (and encouraged!) to perform this type of sensitivity analysis by rerunning the optimization algorithm after varying the inputs that you want to test.

Other types of what-if analysis might involve more strategic or fundamental changes that are different in nature from sensitivity analysis. For example, you may decide to force certain parts of the solution that were previously allowed to change, or you might allow certain things to change that had previously been fixed. You can test specific hypothetical events to see how the solution would change. Suppose, for example, that you were responsible for creating an evacuation plan for a major city in the case of a major natural disaster or terrorist attack. In doing this analysis, you would expect to do many different what-if scenarios surrounding different possible conditions that might exist. The model might be run assuming that public transit was working or not working. You might test how the plan would change based on whether certain highways and bridges were operable or not. For an airline making capital plans, it would be important to perform what-if analysis with different assumptions about the growth of air travel expected in upcoming years.

Because optimization was traditionally the realm of experts who perhaps were more excited by the mathematics than by the application to real-world problems (and their inherent messy nature), we believe there probably has been too much historical emphasis on finding *THE ANSWER*. In truth, no optimization model will ever be perfect. There will be data that is inaccurate. There will be a fine line between "hard" and "soft" constraints. There will be different, often competing, objectives. Moreover, optimization aims to prescribe solutions for future action, and the future is unknown. Therefore, it is highly advised that whenever possible (given time and resource constraints), you exercise the optimization model through sensitivity and what-if analysis. If your entire plan relies on a certain assumption and all heck breaks loose if that assumption fails to hold, it would be much better to know that ahead of time and make alternate plans or have a contingency plan in place than to suffer in the moment. Optimization algorithms are to some degree like a magic black box, but it is up to you to ask the right questions.

Part III
Conclusion

9

Revenue Management

We've highlighted various applications of analytics throughout this book, but we thought that revenue management deserves a chapter to itself. It is a nice topic to cover at the end of the book because it touches on all aspects of analytics—descriptive, predictive, and prescriptive. In addition, it is an important topic to any company that can set prices for its products. Finally, consumers directly see this type of analytics in action as they watch price changes—sometimes without a seemingly good reason.

Revenue management, at the highest level, is about setting the right price to maximize revenue. It is important to note that it is about setting the *right* price—not necessarily a high price. Your total revenue is determined by price multiplied by the amount you sell. If you think back to demand curves from economics, you know that you may be able to maximize revenue by setting a low price and selling a lot or by selling fewer items but at a much higher price. In your Econ 101 class, it probably seemed (at least it did to us) that the solution to every problem could be summarized by saying Marginal Revenue = Marginal Cost. If this were always the answer in the real world, revenue management would not be a particularly interesting field. So, what is different here? Basically, the places where revenue management techniques are very critical violate a couple of important assumptions embedded in those Econ 101 classes. First, for many firms where this is most critical (for example, airlines, hotels, and so on), the marginal cost of each unit is essentially zero or marginal (that is, it costs the airline very little for each additional passenger on a given flight). Second, for these same firms, there is often a fixed quantity available to sell at any one time (that is, airlines can't put more people on a plane than can fit or sell you a ticket for a flight that left yesterday). Finally, those Econ 101 models generally assume that consumers have many choices for an item and are completely "rational" in their choices, which misses the idea that consumers may pay more or less, depending on the reputation of the firm, the perceived quality of the goods, status and luxury considerations, and so on.

Revenue management decisions can often be based on data, thus making revenue management ripe for analytics. More companies can now base their pricing decisions on better data. In fact, this is more and more becoming the case.

With retail pricing available on the web, companies can now monitor the prices their competitors charge and quickly change their prices in response. In many standard consumer goods, there is a big incentive to have the lowest price for an item—and that will often show up at the top of a search for that item. A 2012 *Wall Street Journal* article discussed how some online companies are using sophisticated software to monitor prices and ensure that their product is a penny less than the competitor's.[1] This is a case of simple descriptive analytics—you collect data to describe the prices your competitors are charging and then decide what to do. As a side note, you might have wondered why retailers offer unique versions of many different products; it is to thwart this practice of directly comparing prices.

But, revenue management goes much deeper than just collecting data on your competitor's prices. Like other areas within analytics, the field of revenue management is quite rich, it has been well-researched and studied, and it has been used in practice for a long time.

You can break revenue management into four main categories[2]:

- **Pricing tools**—With pricing tools, you try to determine how the demand changes with changes in price. Or you monitor the price your competitors are charging and adjust accordingly.

- **Forecasting tools**—With forecasting tools, you try to determine what the expected demand is. You can see that this is a part of the demand planning tools discussed in Chapter 6, "Predictive Analytics."

- **Inventory tools**—With inventory tools, you try to determine how much inventory to make available. That is, when you have a limited amount of something to offer (seats at a training seminar, special runs of a watch, and so on), you have to decide exactly how much inventory to offer.

- **Yield management**—Finally, with yield management, you try to decide how much of your fixed inventory (or fixed offering) you should allocate to each of your different customer groups so that you maximize profit. In other words, you get the most "yield" from your fixed inventory. Yield management touches on each of the other tools listed here and provides a good case study with descriptive, predictive, and prescriptive analytics. We'll expand more on this in the rest of the chapter.

Yield management is probably the least intuitive of these revenue management tools. For example, monitoring competitors' pricing and then changing your own price may require good systems and a good process. But the concept is not difficult to grasp: Make sure your prices are lower than your competitors' (or within a range). On the other hand, it may not be well known that the airline industry uses yield management techniques, but it is not hard to find someone who complains about it or is befuddled about the seemingly random price changes for airline tickets. After we explain yield management, you'll understand what is going on. You may still complain that you don't like the price offered, but the small consolation is that you will know what is happening.

In the airline business, the seemingly inexplicable price changes are driven by programmatic mathematical logic. Let's start with a simple example and then expand it. An airline knows that it has a flight between Atlanta and Dallas with a set amount of seats. It knows that a lot of passengers would like to buy a very cheap ticket. And, generally, these customers would be willing to purchase well in advance. But if it takes all these passengers, the airline knows that it will miss the opportunity to sell the tickets for a higher price to business customers who won't buy the tickets until closer to the departure day. So if the plane seats only 125 people, the airline may sell only 75 seats at the low price, and once those seats are gone, raise the price and wait for the 50 business travelers. This explains why you may see a sudden jump in price. The airline now believes that it will have enough customers willing to pay more to fill up the plane.

But, of course, the airlines are more sophisticated than this simple example suggests. They can change prices very frequently. Still, the basic idea remains. At any given point, the airline needs to make a decision on how many seats it should continue to hold for customers willing to pay more. If it expects a larger number of last-minute business travelers, the lower prices may go away faster. If it expects fewer, then the lower prices may stay longer.

American Airlines is known as the pioneer of yield management. And this did not come easy: The company has been working on this since the 1960s. To make this work, American first had to build a system to track all the available seat inventory and the prices paid for each seat. This was not a trivial task, and it pushed the computing power at that time. In analytics terms, this was a descriptive analytics system: American could see all the data and understand the price it was getting for each seat, and it could see how many planes were full. Once it knew the seat inventory, it had to start to build predictive models to understand when customers would purchase tickets and how prices would impact demand. This is predictive analytics.

Finally, American built sophisticated optimization models to help determine what it should do. This is prescriptive analytics. The airlines optimization problem for yield management is quite challenging, but it gives some nice insight into the problem.

The first part of the optimization problem is for the airline to decide how many extra seats to sell. This is called *overbooking*, and it's a practice that drives the public crazy. If a plane has 125 seats, the airline may sell tickets for 140 seats. The reason it does this is that it is virtually certain that for any flight there will be no-shows and cancellations. Predictive models help estimate how many of these to expect for a given flight. In the past, up to 50% of people with reservations would cancel or not show up. American Airlines calculated that if it did not overbook, 15% of its seats would be empty versus 3% on certain flights. Of course, when deciding the number of extra seats, the airline is making a bet on the actual number of people who show up versus how much it will cost the company if it has to bump someone from a flight. Now, some airlines have very strict change or cancelation policies. So, with these in place, they don't need to overbook as much because they will still get revenue from a seat that's empty at flight time.

Once the number of extra seats to sell is determined, the prescriptive models must determine how many seats at each price point to make available. These systems are very dynamic and adjust as they get more information. Many years ago, American Airlines was making 50,000 price changes daily. At this point, the system was doing most of the work automatically and flagging exceptions for managers to look at. Long before we started using the term *machine learning*, the airlines were doing it.

A final note is that it isn't clear that a big airline can survive without yield management. A good story about the power of yield management comes from a talk one of the authors of this book heard from someone who was a vice president of People Express in the late 1980s. People Express built a business of offering single low-cost fares on popular point-to-point routes (avoiding the hub-and-spoke structures of the large airlines). The People Express business model would work only if the company could sell out all its seats; the ticket prices weren't high enough to cover costs if they didn't do this. This former vice president recalled the panic when a big airline (if our memory is correct, it was American Airlines) started using better yield management on the routes that competed with People Express. The big airline could offer a certain number of very low-priced tickets (to fill up seats that would otherwise be empty) and leave enough seats for business travelers (who would bring in enough revenue to make the flight profitable). This had the impact of siphoning off passengers from People Express. And with fewer passengers and more empty seats, there was no way to stay in business. Besides showing the power of yield management, it also shows the power of analytics: If you can use data creatively and with powerful tools, you can change an industry. (Of course, the industry continues to evolve. Southwest Airlines grew quite large without using yield management. Currently several low-cost airlines in Europe seem to be thriving without yield management but by charging for all kinds of services on the plane, such as choosing a seat, getting beverages, checking luggage, and so on.)

Of course, you might think that Southwest offers a counterexample. Southwest Airlines grew into a large carrier by basically having a single low price for point-to-point routes. But it is interesting that as the company has grown larger, its pricing structure is closer to yield management practices, with multiple tiers of pricing. As we become more enamored with analytics, it is always important to remember examples like Southwest, where strategy and good human resource practices are also used to build successful businesses.

As a side note, the airlines need to constantly adjust their yield management strategies. The book *Big Data* references a story about a company called Farecast (which is now part of Bing). Farecast's founder was frustrated with the constantly changing prices of the airlines and realized that he could gather all the prices and all the price changes on airline routes. When airlines first started yield management, they had the upper hand because it was hard to track the price changes. Farecast could pick up all these changes. It would then advise customers on a given route whether the price they were quoted was likely to go up or down if they waited. In effect, Farecast was throwing a wrench into the optimization models of the airlines. The book *Big Data* uses this as an example of how data availability can change an industry and shift the power. We were a bit disappointed that *Big Data* stopped with just that story. We can't believe the airlines would just sit idly by. They could also access information about whether fares are likely to go up or down and adjust their predictive models accordingly. It is an interesting analytics arms race.

Yield management started with the airlines but is now rapidly expanding to other areas, such as hotels, casinos, rental cars, cruise ships, and many other industries. The press is starting to report on some of these stories. When the press covers yield management, it often uses the term *dynamic pricing* to describe it. The *Chicago Tribune* reported that the popular Goodman Theater (Chicago's oldest and largest nonprofit theater) was moving to dynamic pricing. That is, seats on the more popular nights will be priced higher. If seats for a given show aren't selling well, Goodman will lower the price. We recently saw that the University of Michigan will apply dynamic prices for football tickets. We aren't sure how it will do this, but this could be a chance for the school to capture some of the profit that goes to the scalpers of the Ohio State game. *National Geographic* magazine reported that parking meters in San Francisco are moving to dynamic pricing. After a little research, it seemed like dynamic, in this case, meant changing prices only about once a month. But the fascinating thing was that something like dynamic pricing was mentioned in *National Geographic*; it seems like it has caught the attention of the public.

So, how can you determine where yield management will work well? There are several factors to note here:

- **Limited or perishable products**—Another way to think about this is that these industries deal in products for which it is impossible to stock inventory. For example, once an airplane takes off, an empty seat can no longer be sold. Once a show begins, the empty seats have no value. Once a night passes, an empty hotel room has no value. The list goes on, and creative new examples pop up every day. Airbnb realized that an empty room in your house could be rented.

- **Commitment required when future demand is unknown**—If an airline offers cheap tickets now and sells too many of them, there won't be enough seats available when the customers willing to pay more start to show up. Conversely, if it doesn't sell enough of the low-priced tickets, the plane will have empty seats if not enough business travelers show up.

- **Some way of segmenting customers**—Different customer segments are willing to pay different amounts. Yield management strategies are most useful when you can identify those segments. In the case of the airlines, they may not have a direct way to identify business versus leisure customers (although we wonder if they can discern this at least when a reservation is made from a corporate travel site), but most times they can make a very good prediction based on time of booking. It is pretty unusual for people to plan leisure travel at the last moment (although airlines try to capture that market as well, with last-minute weekend fares).

- **Satisfying demand from many customer groups with a unit of product**—Yield management works if you have some flexibility in your product. For example, the same hotel room can be rented to someone who books months in advance for a great rate, someone who is getting a discount because she is part of a wedding party that booked a block of many rooms, or a business traveler who pays a premium price because he booked a day in advance.

If your problem has characteristics like these, you may have an opportunity to use the well-developed techniques of yield management.

We'll wrap up this chapter with a story from Harrah's casinos. Harrah's was highlighted in Davenport and Harris's book *Competing on Analytics*. Harrah's is sophisticated when it comes to analytics. It is no surprise that it is good at yield management.

An *Interfaces* article conveys a story of one of the Harrah's casino properties in rural North Carolina (about a three-hour drive from Atlanta) with an onsite hotel.[3] The hotel has a creative way to implement yield management. As background, Harrah's has a very extensive loyalty card system that allows the company to track the details of each of its customers. And their customers have an incentive to use the card because it pays them nice rewards. So Harrah's knows a lot about its customers. At this particular hotel, Harrah's knows that more people will show up on weekends than on weekdays, and it has good predictive models that help it determine when people are likely to call in for a reservation at the hotel. A customer may call up on Friday morning for a Friday night stay, and Harrah's may tell the customer it is full but agree to pay for the customer to stay at a hotel up the road. Presumably, the hotel up the road is not as convenient for the customer, but Harrah's could agree to pay for it. The surprising thing is that there may be plenty of rooms available at the onsite hotel. Harrah's is holding out for other customers, customers it expects will be willing to pay more. Despite these expectations, Harrah's will not actually charge more for the room for this second group. Instead, Harrah's is using the time of booking request as a means to segment customers by how much they spend, in total, at the casino. Harrah's has forecasted how many high rollers are likely to ask for reservations on Friday afternoon, and it saves rooms for them. Taking this a step further, Harrah's may even charge these customers *less* for the hotel than other guests would pay. The reason it does this is probably clear to those familiar with the gaming industry. Harrah's is not really in the hotel business but rather in the gambling business. It generates more total revenue by having big gamblers stay at its hotel (even at a reduced room rate) than it generates from people who stay at the hotel but spend a few hours playing nickel slot machines before eating at the buffet.

There are two good points to highlight with this story. The first is that Harrah's can segment its customers like this because its loyalty card system is so good. This shows the value of loyalty cards and understanding customers.

The second point is that, in this case, the interests of the best customers and Harrah's seem to be aligned. The best customers (in Harrah's terms, those who gamble the most) get the best perks (saving spots at the hotel and good prices for the room). Contrast this a bit with the airline industry. The best airline customers are the business and last-minute travelers; they pay the most and make the route profitable. But, presumably, these customers do not necessarily want to pay so much more than their fellow travelers. Yes, they get a few perks like boarding first, maybe getting better seats, and being able to make last-minute reservations. But in the end, they still have to pay more. This is unlike in the casinos, where the best customers are willingly spending their money as a leisure activity.

10

Final Tips for Implementing Analytics

Our opinion, based on professional experience, is that when it comes to analytics, organizations focus far too much on the tools (software) and too little on who will be using the tools. In many cases, companies will purchase a sophisticated and expensive software package and then deploy some of their least experienced people as the primary users of these tools.

It Is the Archer, Not the Arrow

Both of the authors of this book have fathers who can build and fix things around the house. One of them is a professional carpenter. Neither author of this book picked up those practical skills, and they are both quite useless at fixing problems around the house. Now, let's say we needed to build a hand railing for a staircase. If you gave our fathers the most primitive of tools, they would do a much better and much faster job at building this hand railing than we would, even if we had the latest and fanciest tools available. In fact, even with great new tools, we might not even be able to complete the job or we might create a hand rail that would immediately collapse when someone used it.

The same analogy holds for the analytics tools discussed throughout this book. We've discussed many new and fancy tools. However, you can't simply expect that these tools will automatically work. You need to make sure to put these tools into the hands of people who can use them—people who are comfortable with data, who can translate the results from tools to business problems, and who can spot minor problems and make adjustments.

If you want to successfully implement these tools, you need to give them to the right people and provide the proper training for them. Many firms will find different ways to organize their teams to make this happen. But, to be successful, you need analysts who understand the tools, are willing to learn new tools, and understand the theory behind the tools. The analysts also have to be comfortable working with data sets (some large), willing to let the data take them wherever it does, and be able to translate the results to solve business problems. Your first-line analytics managers don't necessarily need to know the details of how the tools work, but they should know the theory behind the tools, have a sense of what tools work in different situations, and know about the common pitfalls and traps. They should be even more capable of translating the results into business decisions and be able to communicate this throughout the organization. Upper management and others in the organization should have some basic understanding of the field of analytics, be willing to ask tough questions of the analytics team, and be willing to learn new things. The data may prove some of your thinking wrong, or you might need to learn something about a technical field because it changes how you do business.

One of the reasons that analytics is a hot topic is because there are a lot of people who are enthusiastic about learning more about it and applying it to solve important problems. Your organization should tap into that enthusiasm.

Wrapping Up

At the core, analytics is about making better decisions and gaining deeper understanding with data. To give it more meaning and to better understand the type of output you can expect, it is best to think about analytics in terms of descriptive, predictive, and prescriptive categories.

Within each of these categories are many different tools you can use. This book has given you a flavor of the tools that are out there and a starting point for better understanding what is best for your business. It is good to remember that we may have ignored a tool or given a tool only a few paragraphs. There are too many tools to cover in a book like this. And there are many books or programs of study that focus on just a single tool.

Also, new tools will continually be developed. The newest of these tools are being grouped under a term called *machine learning*. Tomorrow, there may be another set of tools, and the term *machine learning* may be out of favor. But, the current machine learning algorithms will likely continue to be a valuable part of analytics—just like regression analysis continues to be a powerful tool in analytics.

The availability of data is driving analytics and has even created the new term *Big Data*. Although some people are using the term *Big Data* to refer to the entire field of analytics, all serious discussions around data recognize that you have to do something with the data to make it worthwhile. And that "something" is descriptive, predictive, or prescriptive analytics.

A very positive trend coming out the field of analytics is the desire to test ideas with data in a rigorous manner—like A/B testing of website design instead of guessing or discussing the best course of action. We think this trend helps drive the field of analytics further: More managers want to test ideas with data that turns more and more to the field of analytics for answers.

We've given you a good tour of the field of analytics. It will continue to evolve and change, but the fundamentals will likely stay stable for a long time.

Nontraditional Bibliography and Further Reading

As you've seen, the field of analytics is quite broad and deep, and it is moving quickly. There are a lot of great books and articles available to help you explore different areas in more depths. While we were researching for this book, we came across a lot of good material.

We wanted to break from the traditional bibliography and not only list the books and articles we used as references but to write a sentence or two about the key ones to give you more context. This is not meant to be a book review—but you can assume that we would recommend each of the books on this list for certain purposes. It is meant to be nice reading list for you.

We will also maintain and add to this list and the links in the endnotes on the book's website: www.managerialanalytics.com.

Blackett's War: The Men Who Defeated the Nazi U-Boats and Brought Science to the Art of Warfare, by Stephen Budiansk (2013). If you like analytics and history, this is a good book for you. We used some stories from this book to highlight some key points. Also, some have claimed that the groups of scientists working for the Allies provide the first examples of analytics and the first use of Big Data (although not called that at the time). In addition, this book gives a nice account of trying to use analytics in an environment rife with politics.

The Functional Art: An Introduction to Information Graphics and Visualization, by Alberto Cairo (2012). In Chapter 6, "Predictive Analytics," we mentioned that visualizations shouldn't become art. This book provides some good principles for making sure your visualizations are conveying good information.

Analytics Lessons Learned (www.leananalyticsbook.com) by Alistair Croll and Ben Yoskovitz (2013). This is a free e-book that is a companion to the same author's book *Lean Analytics*. The e-book is just a collection of cases, but it is very well done and gives you a good selection of different analytics solutions.

Competing on Analytics: The New Science of Winning, by Thomas H. Davenport and Jeanne G. Harris (2007). We give this book and the *Harvard Business Review* article of the same name a lot of credit for kicking off the analytics movement. This book was written to show how firms could compete and win by using analytics. It provides case studies and talks about how a company can change its culture to be more focused on analytics. The article, "Competing on Analytics," that proceeded the book, was published by the journal *Harvard Business Review* in January 2006 and written by Thomas H. Davenport. In our view, this article really got the analytics movement going. We have seen it frequently mentioned in the most influential *Harvard Business Review* articles of all time.

Taming the Big Data Tidal Wave, by Bill Franks (2012). This book for the general business reader has a nice section that defines and gives examples of Big Data. It also ties Big Data back to analytics tools. Finally, it discusses how you implement these ideas within an organization.

Machine Learning in Action, by Peter Harrington (2012). Although this book was written for people who want to code machine learning algorithms in the programming language Python, you can read the book and skim the Python code. The book provides a nice framework for thinking about the different types of machine learning algorithms, provides an overview of some of the key algorithms (with enough depth to understand how they work), and provides some nice examples.

The Elements of Statistical Learning: Data Mining, Inference, and Prediction, by Trevor Hastie, Robert Tibshirani, and Jerome Friedman (2009). This is a key textbook on predictive analytics. The book is technical. The first few chapters provide some nice examples. If you have some previous background (from the book you are reading now and other experiences), the examples may give you some additional insight. If you really want to learn the technical details, get this book.

A Tour through the Visualization Zoo, by Jeffrey Heer, Michael Bostock, Vadim Ogievetsky (May 1, 2010). This is a handy guide to the purposes of different types of charts.

Managerial Statistics: A Case-Based Approach, by Peter Klibanoff, Alvaro Sandroni, Boaz Moselle, and Brett Saraniti (2006). This is a textbook for business students written by professors at Northwestern's Kellogg School of Management. This textbook is a great book for helping you learn applied statistics. The book comes with an Excel add-on that lets you practice the concepts.

Moneyball: The Art of Winning an Unfair Game, by Michael Lewis (2004). This book presents a great case for the use of descriptive and predictive analytics in baseball. This book helped popularize the use of analytics.

Uncontrolled: The Surprising Payoff of Trial-and-Error for Business, Politics, and Society, by Jim Manzi (2012). This book is wide in scope, going well beyond business, but we focused on the sections of the book devoted to business. The book opens with a detailed and well-researched discussion of the scientific method and how we can use it to show something to be true. It is important for people working in analytics to understand the scientific method. Then, the book discusses the importance of randomized field trials and why they are important to business. This idea of testing is important to the analytics movement.

Big Data: A Revolution That Will Transform How We Live, Work, and Think, by Viktor Mayer-Schonberger and Kenneth Cukier (2013). This book received a lot of press, and we present many ideas from it in this book. It gives a creative definition of Big Data and should give you some creative ideas for your business. It also discusses some of the social and political implications of Big Data that are beyond the scope of this book.

Handbook of Statistical Analysis and Data Mining Applications, by Robert Nisbet, John Elder IV, and Gary Miner (2009). This is an extensive textbook on data mining. We were particularly inspired by the chapter on the top 10 data mining mistakes.

Innumeracy: Mathematical Illiteracy and its Consequences, by John Allen Paulos (1988). This short book is a classic on thinking about numbers in a logical way.

The Optimization Edge: Reinventing Decision Making to Maximize All Your Company's Assets, by Steve Sashihara (2011). This book is meant to introduce optimization (prescriptive analytics) to the general business reader. It does so with plenty of cases and examples. The examples cover a wide range of applications, so it should help give you ideas on where optimization could help you. It also devotes large sections to how to implement these solutions. The section on implementation applies equally well to any type of analytics solution.

The Flaw of Averages: Why We Underestimate Risk in the Face of Uncertainty, by Sam Savage (2009). This book explains why you can get into trouble if you just use simple averages to make decisions. If you are going to be good at analyzing data and building models, you need to have a good grasp of variability and how it can impact a decision.

A First Encounter with Machine Learning. By Max Welling (www.ics.uci.edu/~welling/teaching/ICS273Afall11/IntroMLBook.pdf, 2011). We primarily used Chapters 3 and 4 of this book to cover the general idea of learning and the types of machine learning. These chapters have a nice explanation of over- and under-fitting your models.

Naked Statistics: Stripping the Dread from the Data, by Charles Wheelan (2013). This is a good introduction or reminder of statistics for the general population. If you don't remember anything about statistics, this is a good starting point. Even if you know statistics, you will probably find this book helpful when you have to explain to others. It provides a good reminder that the field of statistics is important to the field of analytics.

A Programmer's Guide to Data Mining: The Ancient Art of the Numerati, by Ron Zacharski (http://guidetodatamining.com, 2012–2013). Although this book is written for people who want to program machine learning algorithms in Python, we found that the explanations of the algorithms are very intuitive. For example, this book has a good write-up on recommendation systems. You could follow along with the calculations in Excel if you didn't want to code the examples in Python.

Harnessing the Power of Big Data: The IBM Big Data Platform, by Paul Zikopoulos, Dirk deRoos, Krishnan Parasuraman, Thomas Deutsch, David Corrigan, and James Giles (2103). This book is about IBM's Big Data platform. Here Big Data refers to the IT definition: a lot of data. The introduction to this book is useful for the general reader. It helps define the technologies around managing large data sets. You may also find related information from IBM by searching on the title of this book.

Endnotes

Chapter 1

1. "GE Tries to Make Its Machines Cool and Connected," *Businessweek*, December 6, 2012,

2. "Data Crunchers Now the Cool Kids on Campus," *The Wall Street Journal*, March 1, 2013. It is easy to find other articles that discuss the demand for people with analytics skills.

3. These stories come from *Competing on Analytics*. See "Nontraditional Bibliography and Further Reading" for details.

4. The INFORMS website (www.informs.org) says: "The Institute for Operations Research and the Management Sciences (INFORMS) is the largest professional society in the world for professionals in the field of operations research (O.R.), management science, and analytics." It points to a lot of valuable information.

5. This image is in the public domain: http://commons.wikimedia.org/wiki/File:Snow-cholera-map-1.jpg. You can also find a modern update of this map at www.theguardian.com/news/datablog/interactive/2013/mar/15/cholera-map-john-snow-recreated. You can see its modern use in fighting dengue fever at http://networkdesignbook.com/visualization-to-help-stop-dengue-fever/.

6. This case study came from the free e-book *Analytics Lessons Learned: Case Studies on the Use of Lean Analytics* by Alistair Croll and Benjamin Yoskovi. This book has many interesting cases. It is the companion to the book *Lean Analytics: Use Analytics to Build a Better Startup Faster*. The Circle of Moms website is www.circleofmoms.com.

7. The details of this case come from an Optimizely video published on YouTube called "Best Practices & Lessons Learned from 30,000 A/B and Multivariate Tests" given by Dan Siroker, the co-founder and CEO. This link is www.youtube.com/watch?v=7xV7dlwMChc.

8. See "Nontraditional Bibliography and Further Reading" for details on the book *The Optimization Edge*. Also, Steve Sashihara, the author, uses the story of the GPS when he gives talks on his book.

9. The information for the kidney donation came from a presentation at an ILOG (now part of IBM) users group meeting (see http://rcdn-3.brainsonic. com/c1/cdn/3/customers/ilog/20090203_dialog09/large_player/d2_r2_7.html), from a short story that was run on NBC (see www-304.ibm.com/connections/ blogs/ILOGoemPartners/entry/using_ibm_ilog_cplex_to_saves_lives_with_ smarter_healthcare1?lang=en_us), and from an academic paper ("Efficient Kidney Exchange: Coincidence of Wants in Markets with Compatibility-Based Preferences," by Alvin E. Roth, Tayfun Sonmez, and M. Utku Unver in the *American Economic Review*, June 2007).

10. If you have a database of 1,000 donor-recipient pairs, there are about 500,000 different overlapping potential matches to analyze. That is, if pair A is matched with pair B, then A can't be matched with pair C. If you looked at a number of different triple groups, there would 166 million different overlapping groups. And the size will grow much faster than linearly if you add more people to the database (which is good thing to do to help make better matches!).

11. This case study comes from IBM. Here is a link to the sources that helped develop this material: http://optimizationandanalytics.wordpress.com/2013/02/05/ dc-water-and-ibm-a-case-study-in-analytics/.

12. This case comes from a *Businessweek* article, "Coke Engineers Its Orange Juice— With an Algorithm" in the January 31, 2013, issue (see www.businessweek. com/articles/2013-01-31/coke-engineers-its-orange-juice-with-an-algorithm #r=com-s).

13. The details for this case came from an IBM press release: www-03.ibm. com/press/us/en/pressrelease/33290.wss. This case won the Edelman Prize at INFORMS (www.informs.org/Recognize-Excellence/Franz-Edelman-Award), the most prestigious award for applied operations research. There are many good examples of applied analytics on this website. See endnote 5 for more information on INFORMS.

14. The *Competing on Analytics* book and article mention strategy in different ways. The majority of the article talks about analytics as part of an overarching strategy. The book says, "Analytics themselves don't constitute a strategy, but using them to optimize a distinctive business capability certainly constitutes a strategy" (p. 9). The article also gives the example that Allstate has started to "embrace analytics as a strategy." Davenport wasn't trying to define analytics as a strategy, but he has certainly thought a lot about it and adds to this discussion.

15. This quote comes from a nice article on the meaning and types of analytics from four authors from IBM (Irv Lustig, Brenda Dietrich, Christer Johnson, and Christopher Dziekan). *Analytics Magazine* is an online-only publication from INFORMS. This article can be found at http://analytics-magazine.org/ november-december-2010/54-the-analytics-journey.html.

16. See endnote 14.

17. See endnote 14.

Chapter 2

1. The idea is that labor and capital had always been thought of as the two key (and general) economic inputs. Now people were proposing that data be added to this short list. We first read about this notion in a special section of the February 25, 2010, *The Economist* magazine (see www.economist.com/node/15557443). Since then, we have seen this claim in several other places. On August 17, 2013, *The New York Times* published an article by James Glanz titled "Is Big Data a Big Economic Dud?" where he argued that Big Data's impact is not really showing up in the economic indicators, so it may be premature to say that it should be on par with labor and capital.

2. This is from the June 23, 2008, edition of *Wired*. The article that makes this point is called "The End of Theory: The Data Deluge Makes the Scientific Method Obsolete," written by Chris Anderson, the Editor-in-Chief of *Wired*.

3. See "Nontraditional Bibliography and Further Reading" for a short write-up of *Taming the Big Data Tidal Wave* by Bill Franks, published in 2012.

4. Here is the link to Justin Fox's blog October 4, 2010, blog post: http://blogs.hbr.org/fox/2012/10/why-data-will-never-replace-thinking.html.

5. See endnote 1 from Chapter 2. This information came from that article.

6. This case came from the youtube video in endnote 7 from Chapter 1.

7. See "Nontraditional Bibliography and Further Reading" for a short write-up of *Uncontrolled* by Jim Manzi, published in 2012.

Chapter 3

1. We couldn't find the original source of this story, so we are going from memory, and the numbers are for illustrative purposes. The purpose of the story isn't to talk about GE, but to talk about how some businesses are very difficult to forecast, and you shouldn't pretend that a forecast is going to be reliable.

2. Another example of innumeracy could be that the counterfactual is not factual. This is often called the *error post hoc ergo propter hoc* (roughly "after this, therefore because of this") *fallacy*. Manzi points out this fallacy in his book. If you did A and the outcome was bad, it doesn't mean that if you had done B, the outcome would have been better. As an example, we're pretty sure because it's ingrained in all Chicago Bulls fans that Doug Collins was the coach to get the Bulls in position, but they needed Phil Jackson to go take it to the next level (as the Bulls said to justify the move). Since Phil went on to win six titles with the Bulls, many conclude that the Bulls were right and indeed firing Collins was the right move. Well, it may have been the right move, but to conclude this based on Phil winning six titles is to fall to the classic logical error post hoc ergo propter hoc. Yes, Phil won six titles, but are we sure Collins would not also have won six titles? We will never know, but I think most will argue that Collins is one heck of a coach, and the six titles just happened to correspond with Jordan and Pippen coming

into their primes (as well as Grant and Rodman for some of those years). In addition, Bird, Magic, and Isiah were declining...Someone had to win those titles, and betting on the team with the best player in the league seems logical. My guess is that Collins would have also almost certainly won several titles with that team. So were the Bulls right? They weren't wrong, but I'm not sure they were right either.

3. Michael Schrage's book *Serious Play* is devoted to this topic. This book makes the argument that organizations can learn a lot about themselves from the prototypes and models they build.

4. Gary Loveland's quote comes from Davenport's book *Competing on Analytics*.

5. See Michael Schrage's October 29, 2012, blog post, "Why Polling Is Always Political," (http://blogs.hbr.org/schrage/2012/10/the-truth-about-polling-its-always-a-judgment-call.html).

6. The book *Optimization Edge* has a nice list of other biases that can impact our ability to analyze data. These include *mere exposure effect* (just because we know something, we are biased toward it), *outcome bias* (judging by the ultimate outcome, not the evidence you had at the time), *availability heuristic* (bias toward thinking things that can be imagined are more likely to occur), *actor-observer bias* (bias toward attributing others' actions to their personality and your actions to your situation), and *illusory correlation* (assuming that a correlation exists between two events when none does).

Chapter 4

1. The training and test data can be found on the book's website (www.managerialanalytics.com). We used the open source program Weka to run the analysis.

2. We created this decision tree with the open source program Weka.

3. The Pearson coefficient formula can be found in many different formats. We found that the book *A Programmer's Guide to Data Mining* has an easy-to-follow explanation. See "Nontraditional Bibliography and Further Reading" for a short write-up on this book.

4. For a nice explanation of the cosine similarity formula, turn again to *A Programmer's Guide to Data Mining*. See "Nontraditional Bibliography and Further Reading" for a short write-up on this book.

5. This plot was drawn using the open source statistical package called R.

6. Particularly, we recommend *Naked Statistics* and *Managerial Statistics* for more details on how to run regression analysis. See "Nontraditional Bibliography and Further Reading."

7. *Managerial Statistics: A Case-Based Approach*, by Peter Klibanoff, Alvaro Sandroni, Boaz Moselle, and Brett Saraniti (2006).

8. This comes from Michael Trick's blog post "Easy and Hard Problems in Practice," (http://mat.tepper.cmu.edu/blog/?p=1751), which references "David Eppstein of the blog 0xde."

9. For one write-up on this story, see www.theregister.co.uk/2006/08/15/beer_diapers/. You can easily find others with a quick search.

10. Peter Harrington, in his book *Machine Learning in Action*, gives a unique example. His learning data set consists of a variety of characteristics of different mushrooms. In the unsupervised version, the same market-basket type of analysis can help determine what features of mushrooms go together. One data point for each mushroom is whether it is poisonous. In an implementation of a market-basket type of analysis, you can specify that you want to understand what characteristics go with whether a mushroom is poisonous. This is like supervised learning because you are trying to see what other characteristics of a mushroom go with whether it is poisonous. In this case, the researcher wanted to see what characteristics were shared by poisonous mushrooms.

11. Robert Nisbet, John Elder IV, and Gary Miner have a nice chapter in their book *Handbook of Statistical Analysis and Data Mining Applications* on the top 10 data mining mistakes. It may be worth a look if you are embarking on a project.

12. This case is from an article in *The Economist* from February 18, 2012, titled "Mind Your Language."

Chapter 5

1. Information for this case came from the January 22, 2008, *New York Times* article "A Basic Hospital To-Do List Saves Lives" (www.nytimes.com/2008/01/22/health/22brod.html?_r=0) and the December 10, 2010, *New Yorker* article "The Checklist," by Atul Gawande.

2. Recent research suggests that storing data in columns (instead of rows) can speed up certain types of analysis. We should point out that this is a rather advanced approach. For our purposes, if you learn databases in the standard way, you will be in good shape. For more on this topic, see Wikipedia: http://en.wikipedia.org/wiki/Column-oriented_DBMS.

3. This discussion on database basics was inspired by course material developed by Dr. Eitel Lauría of Marist College and posted on IBM's academic initiative. This course does a nice job of covering the basics of databases and provides hands-on exercises. The course information can be found at https://www.ibm.com/developerworks/mydeveloperworks/groups/service/html/communityview?communityUuid=065eaf68-e1e1-409e-9826-75575a1a3d09#fullpageWidgetId=W98e4ea94eba3_4cb4_9c7b_2bacf0bc2e01&file=ef28fa92-7a19-44da-aae1-191290f3e7cc.

4. See the Wikipedia article on Peter Chen for more information: http://en.wikipedia.org/wiki/Peter_Chen.

5. See videos on the Ayasdi's website, www.ayasdi.com, for more information.

6. We show only the first 18 months of data in the table. For the full table, see the book's website (www.managerialanalytics.com).

7. See the book *The Functional Art* (see "Nontraditional Bibliography and Further Reading") for a good discussion of infographics.

8. See Michael Schrage's March 26, 2013, blog post, "The Questions all Smart Visualizations Should Ask," (http://blogs.hbr.org/2013/03/the-question-all-smart-visualizations/).

9. See Bill Frank's March 21, 2013, blog post, "The Value of a Good Visual: Immediacy," (http://blogs.hbr.org/2013/03/the-value-of-a-good-visual-imm/).

10. For a nice example of comparing height instead of area, see the book *The Functional Art* that we discussed in "Nontraditional Bibliography and Further Reading."

Chapter 6

1. See endnote 7, Chapter 1. In addition, see the May 2012 *Wired* magazine article "The A/B Test: Inside the Technology That's Changing the Rules of Business," which also provides nice insight into the rise of A/B testing and its impact on the culture of a company.

2. This is the same *Wired* article mentioned in endnote 1.

Chapter 8

1. We mentioned this contest in endnote 13 for Chapter 1. The link for the cancer research is http://interfaces.journal.informs.org/content/38/1/5.

2. We are appreciative of Irv Lustig and Carol Tretkoff for introducing this term to us. Irv attributes the creation of the term to Lloyd Clark. They created and used this terminology to help educate more people about the power of optimization.

Chapter 9

1. The September 5, 2012, *The Wall Street Journal* article was titled "Coming Soon: Toilet Paper Priced Like Airline Tickets," and it was written by Julia Angwin and Dana Mattioli.

2. Serguei Netessine and Robert Shumsky published a short white paper called "Introduction to the Theory and Practice of Yield Management," in *INFORMS Transactions on Education* 3(1): 34–44, September 2002. This paper is a good resource for learning more about yield management and seeing a sample problem.

3. The material for this story came from *Interfaces*, 38(3): 161–175, May–June 2008, by Richard Metters, Carrie Queenan, Mark Ferguson, Laura Harrison, Jon Higbie, Stan Ward, Bruce Barfield, Tammy Farley, H. Ahmet Kuyumcu, and Amar Duggasani.

Index

Numbers

3D pie charts, 133
80/20 rule, 45-46

A

A/B testing, 143-146
accounting data, 47-49
Airbnb, 33
airlines, yield management, 205
alerts, descriptive analytics, 124-126
algorithms
  fast algorithms, 190
  genetic algorithms, 194
  k-means algorithm, 83-87
  machine learning, 67-69
    supervised learning. See *supervised learning*
  market-basket algorithms, 87-89
  optimization, 187-188
    approximation algorithms, 190-192
    exact algorithms, 188-190
    heuristics, 192-193
    what-if analysis, 195-197

  simulated annealing algorithms, 194-195
  tabu search algorithms, 195
Amazon, predictive analytics, 9
American Airlines, yield management, 204
analytics
  competing on, 20-21
  defined, 5-7
  descriptive analytics, 7-9
  managerial analytics, 17-19
  new features, 13-14
  overview, 3-5
  predictive analytics, 9-10
  prescriptive analytics, 10-11
  types of, 15-17
  without Big Data, 32
analytics mindset, 42-44
  80/20 rule, 45-46
  accounting data, 47-49
  incorporating variability into analysis, 46-47
analytics models, 42-43
approximation algorithms, 190-192